时尚设计
FASHION DESIGN

（中英双语版）

[法] 戴安娜·代尔瓦勒 著
by Diana Derval

案　例　cases
练　习　exercises
游　戏　games

東華大學出版社
·上海·

图书在版编目(CIP)数据

时尚设计 / （法）戴安娜·代尔瓦勒著. -- 上海：东华大学出版社，2024.6

ISBN 978-7-5669-2359-2

I. ①时… II. ①戴… III. ①服装设计－研究 IV. ①TS941.2

中国国家版本馆CIP数据核字（2024）第082061号

责任编辑　吴川灵
封面设计　雅　风

时尚设计
SHISHANG SHEJI
[法]戴安娜·代尔瓦勒　著
东华大学出版社出版
（上海延安西路1882号　邮政编码：200051）
新华书店上海发行所发行　上海颛辉印刷厂有限公司印刷
出版社官网：http://dhupress.dhu.edu.cn/
出版社邮箱：dhupress@dhu.edu.cn
发行电话：021-62373056
开本: 889 mm×1194 mm　1/32　印张: 5　字数: 288千字
2024年6月第1版　2024年6月第1次印刷
ISBN 978-7-5669-2359-2
定　价：58.00元

序 Foreword

时尚产业以其独特的魅力和创新精神，不断地推动着社会文化的前进。作为跨学科、跨领域、跨文化的综合体，时尚融合了艺术、设计、商业、科技等多领域的知识，跨越了不同文化和地域的界限。时尚设计师不仅需要拥有艺术家的直觉与企业家的精神，还需要有科学家的严谨。通过跨学科的探索和实践，创造出兼具有美学与实用价值的时尚产品。

对于时尚产业而言，瞬息万变的动态演进正是产业活力的来源。黛安娜教授与DervalResearch代戈长期致力于以实践与研究来探索时尚行业的创新方法。他们以持之以恒的研究精神，深入洞悉、探究与解决时尚品牌与产业发展中的创新问题。

本书系统介绍了多维度感官剖析与体验转换的时尚设计方法，并以丰富而有趣的实战案例诠释了时尚创新的最佳实践，不仅是对时尚设计领域的一次深刻探索，更是对创新精神的一次精彩诠释。我相信，无论是设计师、学者还是对时尚感兴趣的普通读者，都能从中获得新的启发和灵感。期待本书的创新方法能为您真正赋能，并让时尚更好地服务于我们的美好生活建设！

东华大学上海国际时尚创意学院院长、博士 李峻

In the realm of fashion, its distinctive allure and pioneering ethos are the relentless engines propelling the advancement of societal culture. Fashion stands as a multidisciplinary nexus, seamlessly integrating expertise from the spheres of art, design, commerce, and technology, transcending the confines of culture and geography. The role of a fashion designer encompasses the creative intuition of an artist, the entrepreneurial drive, and the meticulous precision of a scientist. Through a fusion of interdisciplinary studies and practical endeavors, they craft fashion with harmonious blend of aesthetic allure and functional utility.

The dynamic flux of the fashion industry is its lifeblood, a testament to its ceaseless vitality. Professor Diana and DervalResearch代戈 have been steadfast in their commitment to uncovering innovative methodologies within the fashion sector through a blend of hands-on experience and scholarly inquiry. Their unwavering dedication to research has enabled them to profoundly comprehend, investigate, and address the innovative challenges faced by fashion brands and fashion industry.

This book meticulously delineates the methodologies of multi-dimensional sensory profile and experience transformation in fashion design, elucidating the quintessence of innovation in fashion through a plethora of rich and captivating case studies. It represents not merely an in-depth foray into the domain of fashion design but also a splendid tribute to the spirit of innovation. It is my conviction that this book will serve as a wellspring of inspiration and insight for designers, academics, and all aficionados of fashion. We eagerly anticipate that the innovative approaches presented herein will not only empower you but also contribute to the enhancement of fashion's role in enriching our collective pursuit of a more beautiful life.

Dr. Jun Li
Dean of Shanghai International College of Fashion and Innovation, Donghua University

简介 Introduction

当我喝着咖啡环顾四周时 —— 我最喜欢的活动！ —— 我一直在想，“这个女孩为什么穿高跟鞋？这家伙为什么开黑车？为什么法国国旗有三种颜色？是什么让这个设计如此成功？”我一直在研究，直到找到答案。

通过这本书，我很高兴能和大家分享那些没人敢问的问题的答案 —— 也许除了一些烦人的孩子 —— 以便最终解读人们的行为和偏好。

通过测试、商业案例和证明，您将发现：

• 红牛罐头在中国是金色的，在澳大利亚是蓝色的，这是有原因的
• 高跟鞋是为波斯骑士设计的
• 蚊子喜欢某些类型的汽车
• 有时没有颜色是正确的颜色
• 黑色使瓶子看起来更重
• 一些充满爱心的妈妈仍然更喜欢最新的名牌包而不是孩子们的艺术作品
•Dior J'Adore 和 Kenzo 的 Flower by Kenzo 在法国和中国的表现截然不同
• 颜色偏好与我们的视觉有关
• 丝绸对每个人来说都不柔软
• 女汉子富含雄激素
•“跳出框框思考”这一说法真正植根于神经科学
• 黑车和鲁布托鞋 Louboutin 吸引着同样的人

When I have a coffee and look around me – my favorite activity! – I keep on wondering "Why is this girl wearing high heels? Why is this guy driving a black car? How come the French flag has 3 colors? What makes this design so successful?" and I research until I find the answer.

Through this book, I am stoked to share with you the answers to questions nobody dared to ask anymore – except maybe some annoying kids – in order to finally decode people's behavior and preferences.

Along the tests, business cases, and testimonials, you will discover that:

• There is a good reason why Red Bull cans are golden in China and blue in Australia
• High heels were designed for Persian Knights
• Mosquitos prefer certain types of cars
• Sometimes no color is the right color
• Black makes a bottle look heavier
• Some loving mums still prefer the latest designer handbag to their kids' artwork
• Dior J'Adore and Flower by Kenzo perform very differently in France and in China
• Color preferences are linked to our vision
• Silk is not soft for everybody
• Nuhanzi are loaded with testosterone
• The expression "Thinking outside the box" is truly rooted in neurosciences
• Black cars and Louboutin shoes attract the same people

您可能还会发现，让设计更具可持续性的一个简单方法是更好地满足和预测客户的需求，因为这样做会自动限制退货和浪费。

在书中我涵盖了插图，词汇表，进一步阅读，外卖，练习，一个互动游戏，以及模板，你可以下载来提高你的阅读体验。

不要期望收集大量新的数据，相反，你将有机会从不同的角度来看待已知的品牌和情况。开始以更有针对性的方式设计产品、品牌和体验。

我希望这本书和学到的东西会让你大开眼界，迫不及待地想听听你成功的设计！

You might also find out that an easy way to make design more sustainable is to better address and anticipate customers' needs, as doing so automatically limits returns and waste.
I included illustrations, a glossary, further readings, take-aways, exercises, an interactive game, as well as templates you can download to augment your reading experience.
Do not expect to gather a lot of new data but instead you will have a chance to look at known brands and situations from a different angle. And start designing products, brands, and experiences in a more targeted way.
I hope the book, and learnings, will be an eye-opener and can't wait to hear about your successful designs!

Derval

戴安娜 Diana

戴安娜 · 代尔瓦勒教授（图画经由 DervalResearch 代戈提供，作者：Vlad Kolarov）
Prof. Diana Derval (Drawing courtesy of DervalResearch 代戈 , artist: Vlad Kolarov)

致谢 Acknowledgments

每一本书都是一次没齿难忘的冒险：你开始在这里和那里勾勒出一些想法，不知不觉中，你就经营了一家跨国企业！我感到非常幸运，有这么多人愿意为这些章节做贡献，并给他们一个大大的拥抱来感谢他们。

这是我的第二本中文书，也是我出版的第十二本。所以，亲爱的读者们，我想对你们表示感谢，你们的反馈、评论和问题让我继续前进。

行业领袖们的宝贵意见使商业案例变得引人入胜：墨西哥城索菲特改革酒店的 Robert-Jan Woltering 和 Cecilia Vilches，Devialet 的 François Ruault 和 Alix Leblais，Blacksocks 的 Samy Liechti，Moncler 的 Remo Ruffini 和 Roberto Eggs，73 小时的小欠和宇辰，施华洛世奇的 Joan Ng，李宁的吕一禾和 Amor，肖露露、Serena Lin 和 Ken Chen 为 iF 设计中心。

设计师 Enesoe Chan、米其林星级厨师 Corrado Michelazzo、艺术家 Edgar Artis、设计师 Shane Lee、设计师 Xiaojing Huang、室内和景观设计师 TL 和 Bonnie 分享了他们的灵感和交易技巧，以增加学习体验。

如果没有 Johan Bremer、Helen Liu、Sandrine Goldie、Louise Liu、整个 DervalResearch 代戈团队、卡尔蔡司的 Boris Dejonckheere，Imagine Eyes 的 Nicolas Château 和 Loïc Le Kernevez、大阪立命馆大学的北冈 Kitaoka 教授，以及我们尊敬的客户和合作伙伴的支持和奉献，就不可能产生科学发现和测试。

Each book is an unforgettable adventure: you start sketching some ideas here and there and before you know it, you end up running a multinational enterprise! I feel extremely lucky that so many people accepted to contribute to these chapters and thank them all with a big big hug.

This is my second book in Chinese and my twelfth published book. So, I want to show my gratitude to you, dear readers, as your feedback, reviews, and questions keep me going.

The business cases became captivating with the precious input of industry leaders: Robert-Jan Woltering and Cecilia Vilches for Sofitel Reforma Mexico City, François Ruault and Alix Leblais for Devialet, Samy Liechti for Blacksocks, Remo Ruffini and Roberto Eggs for Moncler, Xiao Qian and Cheryl for 73Hours, Joan Ng for Swarovski, Yihe and Amor for Li-Ning, Joe Xiao, Serena Lin, and Ken Chen for iF Design Center.

Designer Enesoe Chan, Michelin star Chef Corrado Michelazzo, artist Edgar Artis, designer Shane Lee, designer Xiaojing Huang, interior and landscape designers TL and Bonnie, accepted to share their inspiration and tricks of trade to augment the learning experience.

The scientific findings and tests would have been impossible to produce without the support and dedication of Johan Bremer, Helen Liu, Sandrine Goldie, Louise Liu and the whole DervalResearch 代戈 team, Boris Dejonckheere for Carl Zeiss Vision, Nicolas Château and Loïc Le Kernevez for Imagine Eyes, Prof. Kitaoka from Ritsumeikan University in Osaka, as well as our esteemed clients and partners.

与我的同龄人和学生的互动 —— 也包括与完全陌生的人的互动！ —— 每天都是我研究的问题来源。我必须感谢他们所有人，特别是上海建桥学院的范教授和学生，上海商学院的王凯茜教授和学生，上海国际时尚创意学院的李院长和学生以及东华大学的学生，IFA 巴黎的让 - 巴蒂斯特安德烈亚尼和工商管理硕士，索邦商学院的卡里姆 · 齐奈伊 (Karim Zinaï)、玛丽 - 伊夫 · 拉波特 (Marie-Eve Laporte) 和 MBA 学生，Yvon Moysan 和 IESEG 参与者、INSEEC 的 Mariem El Euch Maalej 和学生、Christophe Rethore 和 ICN 以及四川农业大学参与者、ESSEC 校友、ISM 校友、复旦大学校友、交通大学校友和 TED 校友。

方方、奥伯里和弗拉德·科拉罗夫用他们才华横溢的插图让这本书充满乐趣。

感谢刘梦鑫（我欠你一个大人情！），黄成悦，安娜贝尔 · 加利亚 (Annabelle Galea)，辛迪，杰西，奥贝里，维克托，丽玛，安妮塔，张羽萱，当然还有我的出版商吴超仪，他们帮助我把我可怜的中文写作变成了一本好书，点铁成金！

我的家人和朋友 —— 在上海、成都、北京、深圳、南京、澳门、香港、东京、巴黎、阿姆斯特丹、埃因霍温、布鲁塞尔、柏林、慕尼黑、法兰克福、苏黎世、米兰、维也纳、迪拜、阿布扎比、多哈、阿比让、亚穆苏克罗、阿克拉、达喀尔、多伦多、墨西哥，和线上 —— 一如既往地给予了我很大的帮助和支持。感谢给我带来第一台电脑的娜塔莉。感谢 Mijn Johan，他给了我这台非常有用的华为平板电脑，最重要的是几个月的爱、鼓励和耐心。

The interactions with my peers and students - but also with total strangers! – are a daily source of questioning feeding my research. I must thank them all, especially Prof. Fan and Shanghai Jian Qiao University students, Prof. Cathy Wang and Shanghai Business School students, Dean Legend Li and Shanghai College for Fashion and Innovation and Donghua University students, Jean-Baptiste Andreani and IFA Paris MBAs, Karim Zinaï and Marie-Eve Laporte and Sorbonne Business School MBAs, Yvon Moysan and IESEG participants, Mariem El Euch Maalej and INSEEC students, Christophe Rethore and ICN and Sichuan Agricultural University participants, ESSEC Alumni, ISM Alumni, Fudan Alumni, Jiatong Alumni, and TED alumni.
FangFang, Aubery, and Vlad Kolarov kept the pages entertaining with their talented illustrations.
Thanks to Helen Liu (I owe you big time!), Buffon Huang, Annabelle Galea, Cindy, Jessie, Aubery, Victor, Rima, Anita, Yuxuan, and of course my publisher Superwu, for helping me turning my poor Chinese writing into a wonderful book!
My family and friends – in Shanghai, Chengdu, Beijing, Shenzhen, Nanjing, Macau, Hong Kong, Tokyo, Paris, Amsterdam Eindhoven, Bruxelles, Berlin, Munich, Frankfurt, Zürich, Milan, Vienna, Dubai, Abu Dhabi, Doha, Abidjan, Yamoussoukro, Accra, Dakar, Toronto, Mexico, and online – were a great help and support as always. Danke to Natalie who brought me my first computer. Bedankt to Mijn Johan who gave me this very useful Huawei tablet I am using right now, and most importantly months of love, encouragements, and patience.

书评 Reviews

Diana 教授撰写的这本书采用了独有的多维度感官方式扩宽了可以深度挖掘消费者行为及偏好需求的方向，更有效结合不同的当地市场提供了实际可操作的具体案例学习，我认为是公司市场部或专业品牌产品营销人不能错过的一本“亮眼”之书——Krystal Wang, Sulwhasoo-Marketing, AMOREPACIFIC

这是一个复杂而充满活力的话题的全面指南。不同角度的独特案例研究，探索品牌如何导航其路线图，启动和执行营销活动以及取得成功的根源。令人印象深刻的图形和工具，促进直观的理解逻辑和方法学的应用。追随作者的脚步既有趣又刺激。——蔡逸玲，Juicy Couture 大中华区总经理

诺贝尔经济学奖得主埃丝特·迪弗洛 (Esther Duflo) 提出这样一个问题，为什么人们宁愿花半年的收入去购买奢侈品呢？恰好戴安娜教授通过这本书回答了这个问题，同时，也让我对于奢侈品之所以能取得成功的因素有更深的了解。——曾泓博 Victor, 上海大学

当我们在讨论品牌时，我们需要充分利用激发受众多重感官灵感的机会，让我们的创造与众不同，这个过程远远超越了“品牌”两字。在本书里，简单的案例和结构就可以帮助我们理解设计与品牌的核心，它总是让创意的过程意义非凡。——Jean Long, 全球峰会创意策划伙伴，FBIF

The book written by Professor Diana uses a unique multi-dimensional sensory approach to broaden the direction that can deeply explore consumer behavior and preference needs, and more effectively combines different local markets to provide practical and actionable case studies. I think it is an “eye-opening” book that companies’ marketing departments and professional brand product marketers can't miss.——Krystal Wang, Sulwhasoo-Marketing, AMOREPACIFIC

A comprehensive guide to this complex and vibrant topic. Distinctive case studies from different angles, exploring the roots of how brands navigate their roadmaps, launch and execute their campaigns, and achieve success. Impressive graphics and tools that promote an intuitive understanding of the logics and methodology applied. It is interesting, stimulating and fun to follow the footsteps of the author.——Charing Choi, General Manager, Juicy Couture Greater China

Nobel laureate economist Esther Duflo asked why people would rather spend half a year earnings on luxury goods. Professor Diana’s book answers the question and gives me a better understanding of what makes luxury products so successful.—— 曾泓博 Victor, Shanghai University

When we talk about brands, we need to take full advantage of the opportunity to inspire multiple senses in our audience and make our creations unique, a process that goes well beyond the word "brand”. In this book, simple cases and structures can help us understand the heart of design and branding; it always makes the creative process meaningful. ——Jean Long, Global Summit Creative Partner, FBIF

对于正在从事或者想成为营销和品牌管理者、视觉营销管理者、店铺经理及销售者的人们而言，这是一本非常有价值且很有帮助的书籍。戴安娜·代尔瓦勒教授对于五感作用的不同视角将丰富并有效地拓展我们对消费者行为和偏好的理解。像她之前的著作一样，这本新书保持了她一贯的高品质，并给读者带来充满活力和愉悦的阅读体验。—— 吴璠博士，上海建桥学院

戴安娜教授善用平实的语句，从创新、感官科学等方面入手，阐述奢侈品是如何像活人一样，牢牢抓住不同兴趣的消费者，如何在竞争激烈的市场中屹立不倒。——黄成悦，新零售专家

这本书以科学数据和理论为基础，辅以案例研究，很容易将你在设计、品牌和消费者之间错综复杂的关系中获得的新知识带入语境。这本书肯定会像吸引好奇的人一样吸引善于分析的人。——Olivier Blanc，艺术总监，高级讲师

这本由戴安娜·代尔瓦勒教授写的书是迄今为止我读过的最好的营销书——这本双语书充满了洞察力、营销分析和案例研究，以增进对感官营销策略的理解。感谢这本书，我们将利用戴安娜·代尔瓦勒教授的见解在我们未来的产品开发上。感谢您为时尚和奢侈品行业的专业人士出版一本必读的书！——李辛西娅，MBA，拉德芳斯创始人

This is a quite valuable and helpful book for people who are working as, or willing to become marketing and brand managers, visual merchandisers, store managers, or sales managers. Prof. Diana Derval's different perspectives on the role of the five senses will enrich our understanding of customers' behaviors and preferences profoundly and effectively. Like her previous books, this new book meets her high-quality standards, as usual, and provides a vivid and pleasant experience for the readers.——Dr. Fan WU, Shanghai Jian Qiao University

Professor Diana is good at using plain words to explain how luxury goods grasp different consumers with different interests through innovation, sensory science and other human aspects, and how to solidify one's position in a competitive market.——Buffon Huang, New Retail Expert

With a gentle touch of scientific data and theory, the book - supported by case studies - easily puts into context the new knowledge you will gain on the intricate relationships between design, brands and consumers. This book will certainly appeal to the analytical minds as much as the curious ones.——Olivier Blanc, Art Director, Senior Lecturer

This book written by Professor Diana Derval is by far THE BEST marketing textbook I've ever read - a bilingual book full of insights, marketing analysis, and case studies to enhance understanding of sensorial marketing strategies. Thanks to this book, we will utilize Professor Diana Derval's insights in our future product development. Thank you for publishing a MUST-READ book for fashion and luxury industry professionals!——Cynthia Lee, MBA, founder of La Defense

目录 Contents

第一章 Chapter 1

找到一个独有风格 Finding a Signature Style

“黑白分明” “It's black and white”

为什么有些人喜欢鲜艳的颜色，而有些人坚持黑色和白色（或金色）？在本章中，我们将通过 Derval 代尔瓦勒色彩测试来了解颜色在观察者眼中是怎样的。创建一个独有风格是至关重要的，我们将学习定义不同的品牌代码索菲特（Sofitel），阿迪达斯 Y-3（Adidas Y-3）和迈克·高仕（Michael Kors）。来自 Enesoe Chan, Yohji Yamamoto 和 Devialet 品牌的小贴士和故事可以帮助你找到灵感。

目录

图 1.1 Aubery 为 DervalResearch 代戈绘制的插画
Figure 1.1 Illustration by Aubery for DervalResearch 代戈

Why do some people love vivid colors while others stick to black and white (or gold)? In this chapter we will discover how colors are in the eye of the beholder, with the Derval Color Test. Creating a signature style is critical and we will learn to define distinct brand codes with the Sofitel, Adidas Y-3, and Michael Kors cases. Tips and stories from Enesoe Chan, Yohji Yamamoto, and the Devialet brand should help you find inspiration.

Content of this chapter

1.1 我们看到多少种颜色？代尔瓦勒色彩测试 How many colors can we see? The Derval Color Test

我是一个时尚评审团的成员，一个年轻有才华的设计师展示了一个精彩的系列。在展览结束时，另一位评委咄咄逼人地问道："为什么整个系列都是黑色的？"在沉默了很长一段时间后，最后设计师回答说："因为我喜欢这样的系列，黑色的。"直截了当的回答让评委团成员抓狂，而我认为这完全有道理，让我们通过 Derval 代尔瓦勒色彩测试来看看为什么。

检查 Derval 代尔瓦勒色彩测试 (图 1.2): 你能算出多少颜色的细微差别？

确保计算你看到的所有颜色的细微差别，例如，如果你看到 3 个细微差别的蓝色，2 个细微差别的绿色和 4 个细微差别的红色，那么总数就是 3+2+4=9。

做好了么？结果见第 20 页。

我们看到的颜色的细微差别取决于我们眼睛中视锥细胞的数量和分布。 所以当我们开始设计时，问题不是我应该选择什么颜色，而是我是否应该使用颜色。对色彩有反应的人会喜欢优衣库或李宁品牌的大胆设计，而对对比和形状有反应的人会更喜欢 Zara、Adidas Y-3 或 Michael Kors 等品牌。

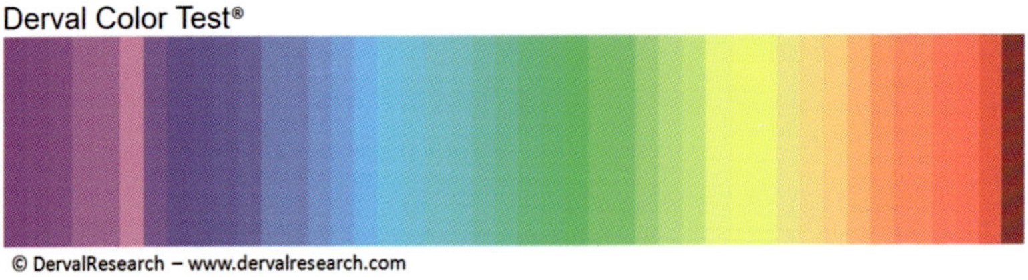

图 1.2 Derval 代尔瓦勒色彩测试（经由 DervalResearch 代戈提供）
Figure 1.2 Derval Color Test (Courtesy of DervalResearch 代戈)

I was member of a fashion jury and a young talented designer presented a splendid collection. At the end of the show, another jury member asked in a rather aggressive way "why is the whole collection black?" There was a long moment of silence, and the designer finally answered: "Because I like the collection like that, in black". The blunt answer got the jury member mad while I was thinking it made total sense and let's see why with the Derval Color Test.

Check the Derval Color Test (Figure 1.2): How many color nuances do you count?

Make sure to count all the color nuances you see, so for instance if you see 3 nuances of blue, 2 nuances of green and 4 nuances of red then total would be 3+2+4=9.

Done? See the results page 20.

The color nuances we see depend on the number and the distribution of cones (=color receptors) in our eye. So, when we start designing, the question is not what color should I choose, but should I use colors at all. People who respond to colors will appreciate UniQlo or Li-Ning bold designs, while people who respond to contrast and shape will prefer the brand codes of Zara, Adidas Y-3, or Michael Kors.

1.2 定义品牌代码：墨西哥城索菲特改革大道酒店 Defining brand codes: Sofitel in Mexico City Reforma

给人以宾至如归感的墨西哥城索菲特改革大道酒店（Sofitel Mexico City Reforma）刚开业，并提供带有法式风味的墨西哥豪华品牌代码（作为更广泛的雅高酒店集团的一部分）。

由墨西哥及中美洲地区副总裁 Robert-Jan Woltering 管理的雅高豪华品牌墨西哥城索菲特改革大道酒店，就是这种独特品牌体验的绝佳例证。历史建筑本身就是周围环境的美丽反映：位于一栋令人印象深刻的 40 层建筑中，该建筑位于改革大道（Paseo de la Reforma）上，这是世界上最具标志性的大街之一，紧挨着象征性的独立天使，墨西哥城索菲特改革大道酒店，正确地将自己介绍为墨西哥传统、法国改良和锐意创新的独特融合。

图 1.3 在改革大道上的墨西哥城索菲特改革大道酒店（经 Sofitel 许可印刷）
Figure 1.3 Paseo de la Reforma, Sofitel Mexico City Reforma (printed with Sofitel permission)

The inviting Sofitel Mexico City Reforma just opened and offers Mexican luxury brand codes with a French twist (as part of the wider Accor hospitality group).

The newly opened Sofitel Mexico City Reforma, under the supervision of Robert-Jan Woltering, as Regional Vice President, Mexico and Central America, ACCOR Luxury brands – is a great illustration of this unique brand experience. The historic structure itself is a beautiful reflection of its surroundings: hosted inside an impressive 40-storey building located on the Paseo de la Reforma, one of the most iconic avenues of the world, and next to the emblematic Angel of Independence, Sofitel Mexico City Reforma rightly introduces itself as a unique blend of Mexican traditions, French refinement, and sharp innovation.

图 1.4 墨西哥城索菲特改革大道酒店绿池（经 Sofitel 许可印刷）
Figure 1.4 Green Pool, Sofitel Mexico City Reforma (printed with Sofitel permission)

图 1.5 墨西哥城索菲特改革大道酒店皇家套房（经 Sofitel 许可印刷）
Figure 1.5 Imperial Suite, Sofitel Mexico City Reforma (printed with Sofitel permission)

作为众多名人的家，这家酒店提供 219 间客房和多达 56 间套房，但最重要的是，一种体验。酒店的品牌代码体现了对细节的关注。例如，蜡烛仪式——灵感来源于路易十四国王统治时期巴黎街道照明的古老习俗——带来了典型的舒适氛围，而索菲特 MyBed™ 的体验，让您从丰富的枕头菜单中选择理想的床上用品，有宾至如归的感觉。餐厅里的爱马仕 (Hermès) 和欧舒丹 (L'Occitane) 设施以及融合的菜肴，以及大使们——每一个团队成员都代表和传达索菲特的体验——将增加法式“手工缝制”的感觉，让你“过法式生活！”

我们热爱（或讨厌）的品牌具有一致和独特的特征，因此我们会随着时间的流逝识别它们。品牌代码有助于弄清品牌的标识，与新的团队成员共享，将其传递给新的设计师，并更容易使品牌保持最新状态 (Derval,2022)。

图 1.6 墨西哥城索菲特改革大道酒店 Cityzen 酒吧（经 Sofitel 许可印刷）
Figure 1.6 Cityzen bar, Sofitel Mexico City Reforma (printed with Sofitel permission)

Home to many celebrities, the hotel offers 219 guest rooms and as many as 56 suites but, most importantly, an experience. The brand codes of the hotel reflect this attention to details. For instance, the candle ritual – inspired by the age-old custom of lighting the streets of Paris during the reign of King Louis XIV – brings a typical cozy atmosphere, while the Sofitel MyBed™ experience, by letting you choose your ideal bedding from an extensive pillow menu, finishes to make you feel at home. The Hermès and L'Occitane amenities and the Fusion cuisine served in the restaurants, as well as the ambassadors – each team member represents and conveys the Sofitel experience – will add the French “cousu-main” touch and let you “live the French way!”
Brands we love (or hate) have something consistent and distinct so that we recognize them over time. The brand codes help clarify the identity of the brand, share it with new team members, pass it on to new designers, and makes it easier to keep the brand current (Derval, 2022).

要建立品牌代码，最容易的就是访问豪华场所、商店或网站。在我们这里，是墨西哥城索菲特改革大道酒店。根据我们的观察，我们可以使用以下多感官描述符来突出其特征：

－颜色：最常用的颜色是什么？什么类型的光？在墨西哥城索菲特改革大道酒店，180° 的视野和建筑让自然光线与非常微妙的天花板建筑照明融合在一起。室内游泳池的绿色墙壁让人“爱死”，并提醒人们只有豪华套房才提供爱马仕的设施。

－材质：使用什么材质或面料？我们可以谈谈休息室沙发的面料。但由于这是一家酒店，也是人们睡觉的地方，索菲特 MyBed™ 的体验值得一提，因为它允许客人在睡眠菜单中选择定制枕头。

图 1.7 墨西哥城索菲特改革大道酒店品牌代码（经 DervalResearch 代戈许可印刷）
Figure 1.7 Sofitel Mexico City Reforma Brand Codes (printed with DervalResearch 代戈 and Sofitel permission)

To build the brand codes, the easiest is to visit the luxury venue, shop, or website. In our case, the Sofitel Mexico City Reforma. Based on our observations, we were able to highlight its features using following multi-sensory descriptors:

- Colors: What colors are mostly used? What type of light? At Sofitel Mexico City Reforma, the 180-degree view and architecture let natural light blend with very subtle and architectural lighting on the ceiling. The living green walls in the indoor swimming pool are “to dive for” and remind of Hermès amenities offered in the luxury suites only.
- Textures: What textures or fabrics are used? We could talk about the fabric of the sofas in the lounge but as it is a hotel and a place where people sleep, Sofitel MyBed™ experience is worth mentioning as it allows the guests to select tailored options of pillows in the Sleep Menu.

– 香味：可以闻到到什么气味或香水？爱马仕的“Eau d 'Orange”带有微妙的橙子香味，深受 Millesime 俱乐部客人的欢迎，而雀巢 Nespresso 咖啡则增加了舒适的感觉。在酒店的公共区域，“索菲特香精”香水正在弥漫。世界各地的索菲特酒店都采用了露西恩·费列罗（Lucien Ferrero）精心打造的标志性香味，它巧妙地融合了清新的佛手柑和泥土般的檀香、清脆的柠檬叶以及白玫瑰、茉莉和山谷百合的甜美花香。

– 象征：什么象征或图标代表品牌？在索菲特，团队成员是品牌大使。索菲特标志象征着一种“文化联系”，这是形成每个酒店的身份和故事的核心。它确保酒店既体现当地文化，又体现索菲特现代法国“生活艺术”。

– 地点：品牌是否指地点？索菲特适应每一个城市，改革大道酒店（Reforma）是如此的墨西哥城！当然也不失法国特色。

图 1.8 墨西哥城索菲特改革大道酒店欧舒丹水疗中心（经 Sofitel 许可印刷）
Figure 1.8 L'Occitane SPA, Sofitel Mexico City Reforma (printed with Sofitel permission)

- Scents: What scents or perfumes are perceptible? The Hermès amenities “Eau d’Orange” with a subtle orange scent is popular among the Club Millésime guests and Nespresso coffee adds to the cosy feeling. In the common areas of the property, the “Essence de Sofitel” perfume is being diffused. The signature scent, used in all the Sofitel hotels around the world and crafted by Lucien Ferrero, is an artful blend of notes of refreshing bergamot and earthy sandalwood, crisp lemon leaf and the sweet floral charm of white rose, jasmine and lily of the valley.

- Symbols: What symbols or icons represent the brand? At Sofitel, the team members are the brand ambassadors. The Sofitel logo symbolizes a “Cultural Link” that is core to forming the identity and story of each hotel. It ensures the hotel reflects both the local culture and the Sofitel Modern French “art de vivre”.

- Location: Does the brand refer to a location? Sofitel adapts to each city and Reforma is so Mexico City! Still, with a French twist.

- 品味：这个品牌传达的是什么味道？特色菜是地中海和墨西哥融合菜——绝对是高级菜。

- 基调：品牌的基调是什么？秉承着“生活无限精彩”的主体，热情是索菲特的关键词。

- 风格：品牌展现出某种特殊的风格吗？当然是法式优雅风。

我们可以添加一个“形状”部分来讨论独特的建筑，在古色古香的房子区域混合了传统的线条，在新建筑中混合了非常现代的线条和几何图形以及多面纹理。

墨西哥城索菲特改革大道酒店设法提供一个故事、建筑、室内设计、豪华服务，是独特的但适合索菲特客人的期望。

图 1.9 墨西哥城索菲特改革大道酒店大堂（经 Sofitel 许可印刷）
Figure 1.9 Lobby Bar, Sofitel Mexico City Reforma (printed with Sofitel permission)

- Taste: What taste does the brand convey? The specialty is the Mediterranean and Mexican fusion cuisine – so definitively Haute Cuisine.
- Mood: What is the mood of the brand? With a motto “Life is Magnifique!”, enthusiasm and “joie de vivre” are the keywords at Sofitel.
- Style: Does the brand show a particular style? French elegance, of course.

We could add a “shape” section to talk also about the distinct architecture, coming with a mix of traditional lines, in the antique house area, and very modern lines and geometric shapes with faceted textures in the new construction.

The Sofitel Mexico City Reforma managed to propose a story, architecture, interior design, and luxury service that are distinct yet fit the expectations of Sofitel guests.

1.3 山本耀司（Yohji Yamamoto）x 阿迪达斯 Y-3
Y-3 by Yohji Yamamoto x Adidas

山本耀司 (Yohji Yamamoto) 以黑色、不对称、简洁、跨性别而又强大的设计创造了一个全球高级定制的标志。

这位著名的时装设计师与阿迪达斯合作创立了运动服装品牌 Y-3。与埃尔顿·约翰、安慰剂乐队和阿迪达斯等客户合作，山本耀司在巴黎和东京获得了许多殊荣，并因其更大的目标而受到赞赏：用时尚语言在中国人和日本人之间架起一座桥梁。

Y-3 品牌是 Y's——山本耀司（Yohji Yamamoto）的原创品牌——自信的高品质运动服装的变体，包括男装、配饰和鞋类，从米兰到北京共有 54 家门店。每个品牌每年产生超过 1 亿美元的收入 (Derval,2016)。

图 1.10 FangFang 为 DervalResearch 代戈绘制的插画
Figure 1.10 Illustration by FangFang for DervalResearch 代戈

Yohji Yamamoto created a worldwide haute-couture signature with black, asymmetrical, simple, gender-fluid, yet powerful designs.
The leading and acclaimed fashion designer started together with Adidas the sportswear brand Y-3. With clients like Elton John, Placebo, or Adidas, Yohji Yamamoto received many distinctions in Paris and Tokyo and is appreciated for his greater purpose: building a bridge between Chinese and Japanese people using the language of fashion.
Y-3 brand is a high quality sportswear variation of Y's – Yohji Yamamoto's original brand – savoir-faire and includes apparel, accessories, and footwear for men and women, with 54 stores from Milan to Beijing. Each brand generates over $100 million revenues per year (Derval, 2016).

Y-3 品牌代码忠于山本耀司最初的价值观。点缀着一抹像红色一样的亮颜色的黑色是品牌的主色调。山本耀司认为，“黑色能同时传递出低调和傲慢，黑色还能表现琢磨不透的慵懒和随性。总之，黑色的宣言就是：我们井水不犯河水”。

山本耀司是不对称设计的先驱。当然，阿迪达斯也忍不住到处添加其著名的条纹。山本耀司解释说，最初他想设计出同样适合女性的男装。

反对快时尚是山本耀司的定位。即使近几年品牌推陈出新的步伐加快了不少，他本人对快时尚和广告的立场依旧不变：“让我像个老年人一样跟你们年轻人谈谈，漂亮的事物日日减少，要小心了，你们没必要买快时尚，特别是年轻人。年轻人本来就因为年轻而自然美丽，你们应该穿简单的牛仔 T 恤，这就够了，不要太时尚。品牌广告常常会让你们为之疯狂。但其实不需要太过性感，你本身就已足够性感。”

虽然 Y-3 尊重了山本耀司的原创价值观，它的风格显然不那么时尚，但仍然很前卫。

在阿迪达斯（Adidas）和耐克（Nike）发起的运动鞋之战中，Y-3 的出现和时尚媒体的曝光发挥了至关重要的作用。在梅赛德斯时装周上，一方面，由于山本耀司的设计，阿迪达斯 Adidas 成功地吸引了 NBA 顶级球星，甚至还有贾斯丁・比伯；另一方面，山本耀司也获得了阿迪达斯（Adidas）脸书上 1500 万名粉丝的关注，可见这完全是一次双赢的合作。

Y-3 brand codes stay true to Yohji Yamamoto’s original values. The main color is black with a pop of bright colors like red. According to Yamamoto: “black is modest and arrogant at the same time. Black is lazy and easy – but mysterious. But above all black says this: I don’t bother you – don’t bother me.”

In terms of shapes, Yamamoto is a pioneer in asymmetric design and of course Adidas couldn’t resist in adding its famous stripes here and there. Yohji Yamamoto explained that initially he wanted to design men outfits that also fit women perfectly.

Yamamoto’s positioning would be anti-fast-fashion. Even though the release of new products has accelerated in recent years, Yohji Yamamoto shares some wisdom about fast-fashion and advertising: “Let me talk like an old man. Young people, be careful. Beautiful things are disappearing every day. Be careful. You don’t need to be shopping at fast-fashion stores, especially young people. They are beautiful naturally, because they are young. So they should even wear simple jeans and a T-shirt. It’s enough. Don’t be too much fashionable. The brand advertising is making you crazy. You don’t need to be too sexy. You are sexy enough.”

Even though Y-3 respects Yohji Yamamoto’s original values, the style is definitively less high fashion but still avant-garde.

In this footwear war opposing Adidas to Nike, Y-3 and fashion media exposure is playing an instrumental role. During Mercedes Fashion Week, Adidas was able to attract, thanks to Yohji Yamamoto’s designs, top NBA stars and even Justin Bieber. In the same way, Yamamoto has access to the 15 million Facebook fans of Adidas: Y-3 is a truly win-win alliance.

1.4 迈克高仕 (Michael Kors) vs 路易威登 (LV)：轻奢的崛起 Michael Kors vs LV: the rise of affordable luxury

迈克高仕（Michael Kors）很快成为了一个非常受欢迎的时尚品牌，其经典而明显的品牌代码可能成为新的 LV。驼色、米色、黑色和可见的金色标志，迈克高仕（Michael Kors）分享了一种喷气机阶层的氛围。你感觉就像在游艇上度过了一个下午。

中国已有超过 100 万名百万富翁——其中数千人被认定为拥有 1 亿人民币或以上财富的超级富豪。他们中有很多居住在北京，其次是广州和上海。中国百万富翁的平均年龄要比西方百万富翁年轻。在活动方面，中国的百万富翁通过私营企业、房地产、股票交易所致富，其中一些是高收入的高管。不出所料，最受欢迎的奢侈品是那些能让你引起注意的：汽车和时尚 (Derval,2016)。

主要的时尚奢侈品买家确实是中国的百万富翁，他们渴望在路易威登（Louis Vuitton）或普拉达（Prada）等欧洲顶级奢侈品牌中展示自己的实力和地位。企业代表使用奢侈品作为企业礼品，尽管最近政府的反腐政策减少了这一部分。而中国中产阶级的数量也在不断增加更微妙的品味，提供新的视角，如迈克高仕（Michael Kors）提出的轻奢。

迈克高仕（Michael Kors），天桥骄子（Project Runway）的明星以及范思哲（Versace）和仰杰（Jimmy Choo）的快乐新主人，创造了手袋历史上最受欢迎的奢侈手袋。小故事是这样的：当 LV 开始推出手包时，它们在巴黎的阿姨（清洁工）中很受欢迎，她们喜欢这种大胆的设计。让我们看看迈克高仕（Michael Kors）能否将其阳光的品牌打造成下一个 LV。

Michael Kors quickly became a very popular fashion brand with classical yet visible brand codes possibly becoming the new LV. With colors like camel, beige, black, and visible golden logos, Michael Kors shares a jet set vibe. You feel like spending an afternoon on a yacht.
China counts over 1 million millionaires – thousands of which are identified as super rich with a fortune of 100 million rmb or more. Many of them live in Beijing, followed by Guangzhou, and Shanghai. Chinese millionaires are on average younger than their Western counterparts. In terms of activities, Chinese millionaires made their fortune from private businesses, real estate, stock exchange, and some are high-earning salaried executives. Unsurprisingly, the most popular luxury items are the ones that help you get noticed: cars and fashion (Derval, 2016).
The main fashion luxury buyers are indeed Chinese millionaires, eager to showcase their power and status with leading European luxury brands like Louis Vuitton or Prada, business representatives using luxury goods for corporate gifting, even though recent government anti-corruption policies reduced that segment, and a growing Chinese middle-class with a more subtle taste, offering new perspectives to brands like Michael Kors proposing affordable luxury.
Michael Kors, star of Project Runway, and happy new owner of Versace and Jimmy Choo, created the most popular luxury handbag in the history of handbags. For a little story, when LV started handbags, they were very popular in Paris among the ayis (cleaning ladies) who loved the bold designs. Let's see if Michael Kors can turn its sunny brand into the next LV.

以贝拉·哈迪德（Bella Hadid）为代表的喷气机阶层品牌以金色的高质量的包袋和高科技而又经典的手表(秘密由 Fossil 品牌制造)而闻名。最新的系列致力于迷人和闪耀的滑行运动。

迈克高仕（Michael Kors）是典型的上班族手袋，这家时装公司用可靠的、时尚的，甚至更实惠的商品说服了从巴黎到上海的中产阶级——供应迈克高仕的车间也为普拉达 (Prada) 工作。

图 1. 11 迈克高仕品牌代码（经由 DervalResearch 代戈提供）
Figure 1.11 Michael Kors Brand Codes (Courtesy of DervalResearch 代戈)

With a slightly jet-set positioning and muses like Bella Hadid, the brand with golden colors is acclaimed for its quality bags and high-tech yet classical watches (secretly manufactured by the Fossil brand). The latest collection is devoted to the glamorous and sparkly sliding sports.
Michael Kors is the typical office worker's handbag, and the fashion house convinced the middle class from Paris to Shanghai with reliable – the workshop supplying Michael Kors is also working for Prada – trendy, and, even better, affordable items.

1.5 与时装设计师 Enesoe Chan 沟通灵感 Channeling inspiration with fashion designer Enesoe Chan

品牌代码是定义时尚身份的关键，贯穿整个时间和空间。正如我们在巴西时装设计师，高级讲师和创意总监 Enesoe Chan 那里看到的那样，他们为新的系列创造过程提供了素材。Chan 用他挑逗、有趣和标志性的风格震撼了从圣保罗到上海的秀场，向我们展示了他的创作过程。

Step 1 – 目的。“时装设计作为一种工具，传达我对社会问题的想法和感受。” Enesoe 解释道，并且他详细阐述道：“我总是有一个明确的目标来设计时装系列。所以我的目标总是与社会问题有关，我从那里获得灵感，如性别、贫困、疯狂、战争、垃圾回收，等等。我用时装设计作为一种工具来质疑、思考和反思社会中的一些价值观。”

图 1.12 Enesoe Chan 的有趣的插图（经由 Enesoe Chan 提供）
Figure 1.12 Playful illustration by Enesoe Chan (courtesy of Enesoe Chan)

Brand codes are key to define a fashion identity, throughout time and space. They feed the new collection creation process as we see with Enesoe Chan, Brazilian Fashion Designer, Senior Lecturer, and Creative Director. Chan, who has shaken runways from Sao Paolo to Shanghai with his provocative, fun, and iconic styles, walks us through his creative process.

Step 1 – Purpose. “I see fashion design as a tool to communicate my thoughts and feelings related to social issues.” explains Enesoe and he elaborates: “I always have a clear PURPOSE for designing the fashion collections. So, my purpose is always related to social issues where I get my inspirations from as gender, poverty, craziness, war, recycling rubbish and so on. I use fashion design as a tool to question, think and rethink about some values in society.”

Step 2－灵感的源泉。Enesoe 将社会事务和环境相联系："关于'垃圾收集者'的时尚是我正走在街上，突然意识到刚刚从我面前走过的垃圾收集者每天步行数英里收集手推车的纸张和塑料，以便出售以回收利用，这可以是我的灵感；上海的时装表演'自行车生命轮'的灵感来自于观察到这么多中国人到处骑自行车，这让我感到惊讶；上海的时装表演' 地面丰富'是关于我发现这座城市有很多源自树木和周围公园的绿色。照片、图像、文字、歌曲、文章、诗歌、名言，以帮助我更好地理解灵感。"

Step 3－颜色、面料和轮廓。Enesoe强调："选择与灵感主题相关意义的颜色。根据季节和质地选择面料来表达主题。考虑衣服的轮廓、图案和具体的剪裁或细节，如拉链、底部、口袋、衬里、线，以更好地显示灵感主题。"

Step 4－心态。Enesoe 证实了"心情正常"的重要性："通常，当我放松并处于平静的心态时，我会得到灵感：散步，小睡，听一些我喜欢的音乐，我在安静、轻便、整洁有序的地方工作得更好，身边有我需要的一切：笔记本电脑、印刷机、扫描仪、手机、杂志、书籍、剪刀、胶水、纸张、彩色铅笔、彩色记号笔。"

Step 2 – Source of inspiration. Enesoe relates to society matters and environment."For 'Rubbish Collectors' I was walking on the street and suddenly I realized the rubbish collectors who had just passed in front of me walking miles a day collecting the cart papers and plastic in order to sell them to be recycled could be my inspiration; the Shanghai 'Cycling Life Wheel' inspiration came from observing so many Chinese people cycling everywhere so I was amazed; the fashion performance in Shanghai 'Ground Richness' was about how green I found the city with a lot of trees and parks around even though being a huge metropolitan city. Photos, images, texts, songs, articles, poems, quotes help me to better understand the inspiration."

Step 3 – Colors, Fabrics and Silhouette. "Choose colors with meanings related to the inspiration topic. Choose fabric according to the season and textures to express the topic. Think about the silhouette, patterns and specific cuts or details on the clothes like zippers, bottoms, pockets, lining, threads to better show the inspiration topic." recommends Enesoe.

Step 4 – Mood. Enesoe confirms the importance of "being in the right mood" and adds: "Usually I get inspired when I am relaxed and in a peaceful mind – walking, taking a nap, listening to some music that I like – and I work better in a quiet, light, clean and organized place with everything I need near me: laptop, printing machine, scanner, mobile, magazines, books, scissors, glue, papers, color pencils, colored markers."

1.6 创造品牌：帝瓦雷（Devialet）令人起鸡皮疙瘩的公司 Inventing a brand: Devialet the goose-bumps company

帝瓦雷（Devialet）是一个相当新的品牌，由三位法国联合创始人创立，他们分别是一位工程师、一位设计师和一位商人，他们对目前音响系统的局限性感到不满。通过帝瓦雷（Devialet）的案例，我们将看到如何运用一点远见和大量创新，不仅可以实现重新洗牌，而且可以改变游戏规则。

用优雅的方式解决问题也是帝瓦雷（Devialet）所做的。这家公司在 10 年内就成为法国工程的标志，它的扬声器和扩音器真的能让你哭。Devialet 的技术能够说服 Jay-Z 和 Bernard Arnault。

除了令人印象深刻的 1.5 亿欧元的融资外，帝瓦雷（Devialet）还成为高保真声音的代名词。在谁将设计出最好的无线扬声器的战争中，现在有了一个无可争议的获胜者，至少对于那些对失真敏感的音乐爱好者而言。帝瓦雷（Devialet）的秘密武器：申请了 200 多项专利，并进行了两次革命 (Derval, 2022)。

图 1.13 帝瓦雷 Devialet Phantom 金版（经 Devialet 许可印刷）
Figure 1.13 Devialet Phantom Gold (printed with Devialet permission)

Devialet is a fairly new brand created by three French co-founders, one engineer, one designer and one businessman, annoyed by the limits of current sound systems. Through the Devialet case, we will see how with a bit of vision and a lot of innovation, it is possible not only to reshuffle the cards but to change the rules of the game.

Solving a problem with elegance is also what Devialet did. The company who became within 10 years the symbol of French engineering proposes loudspeakers and amplifiers that make you cry, literally. Devialet's technology was able to convince Jay-Z and Bernard Arnault.

Beyond the impressive 150 million euros funds raised, Devialet has become synonymous with high-fidelity sound. In the war of who will design the best wireless speaker, there is now an undisputed winner, at least for music lovers sensitive to distortions. Devialet's secret weapon: more than 200 patents filed, and two revolutions (Derval, 2022).

帝瓦雷（Devialet）所进行的第一次革命是在音质水平上，即使在高音量下也能保持水晶般清晰。在帝瓦雷（Devialet）之前，我们的选择只有数字声音和模拟声音。数字声音远不及完美，但具有紧凑而强大的优点。模拟声音最接近人类的声音，但需要更多空间，并且有过热倾向，从而产生失真。帝瓦雷（Devialet）的天才在于将模拟放大器的保真度与 4 个数字放大器的功率相结合，并将其置于一个设计精良的机箱中，在该机箱中，就连铝制螺栓也被铜制螺栓取代，以便更好地调节温度。商业模式的第二次革命得益于一种类似信用卡的放大器（SoC，芯片上系统）。与定位为简单的放大器或智能扬声器不同，帝瓦雷（Devialet）还能够实施合作品牌策略。

竞争对手并不是我们会预想到的。的确，与普遍看法相反，竞争对手并不是生产相似类型产品的公司，而是锁定了您所瞄准的相同顾客钱包的公司。通过改变其在价值链中的地位，帝瓦雷（Devialet）现在能够配备那些曾被认为是竞争品牌的产品。如今，为了显著提高音质，帝瓦雷（Devialet）已经嵌入在许多流行的设备中，如华为智能音箱，它的声学技术也可以推广到智能手表、智能手机、电视或汽车上。

帝瓦雷（Devialet）首席销售和营销官弗朗索瓦 • 鲁欧 (François Ruault) 在负责苹果 iPod 的推出、在微软负责 MR 头显（HoloLens）企业以及最近 Parrot 无人机业务的扩张时，获得了广泛的营销创新知识。他强调道：“音乐是当前的一种载体。它对人有很大的影响——研究正在进行，比如研究它如何调节老年痴呆症患者的认知感知和记忆。”

First revolution operated by Devialet was at the sound level, that stays crystal clear even at high volumes. Before Devialet, we only had the choice between a digital and an analog sound. Digital sound is far from perfect but has the merit of being compact and powerful. Analog sound is the closest to human voice but requires more space and has a tendency to overheat, which creates distortions. Devialet genius consisted in combining the fidelity of an analog amplifier with the power of 4 digital amplifiers, and this inside an extremely well designed chassis, where even aluminum bolts have been replaced by copper bolts in order to better regulate the temperature. Second revolution was in the business model thanks to an amplifier that fits in the equivalent of a credit card (SoC, system on a chip). Instead of Positioning as a simple amplifier or smart speaker, Devialet is also able to implement an “Intel inside” strategy.

Competitors are not the ones we would think of. Indeed, contrary to popular belief, competitors are not companies producing similar types of products but companies targeting the same customer’s wallet you have an eye on. By altering its position in the value chain, Devialet is now able to equip what would have been considered as competing brands. Embedded today in many popular devices like Huawei smartspeakers in order to significantly improve sound quality, the Devialet acoustic expertise can also be rolled out into smart watches, smartphones, TVs, or cars.

François Ruault, Devialet Chief Sales and Marketing Officer – who gained extensive knowledge on marketing innovation while in charge of iPod introduction at Apple, HoloLens enterprise at Microsoft, and more recently Drones expansion at Parrot – highlights: “Music is a carrier current. It has a strong impact on people - research is even being conducted on how it modulates cognitive perception and memory in Alzheimer for instance.”

这种声音的多感官维度已经通过一个意想不到的但很好的组合得到了进一步的探索："Devialet 和 Krug 发布了一个豪华的盒子，将 3 张 Ella Fitzgerald 的唱片（她 1961 年在阿姆斯特丹的 Concertgebouw 演唱会上的表演）与 3 张 Krug Grande Cuvée 的'丢失'版本配对。"弗朗索瓦解释道："我非常喜欢 Krug 的首席执行官奥利维尔·克鲁格（Olivier Krug）将香槟酒比喻为不同葡萄品种的混合酒和音乐的搭配。"

您可以在巴黎、法兰克福或上海新天地的专卖店体验到帝瓦雷（Devialet）的声音，只需在舒适的家中，在 Devialet Phantom 上，或佩戴最近推出的 Devialet Gemini true 无线耳塞即可。在巴黎 Opéra Garnier 的帝瓦雷（Devialet）区域是最独特的，18 个帝瓦雷（Devialet）魅影扬声器加入他们的和声，演绎出一首原声交响乐。"数量和深度是两回事"，弗朗索瓦和帝瓦雷这样总结道。有些人称之为"起鸡皮疙瘩的公司"正在迅速扩张，但总是努力保持灵活性和客户友好度。最好的例子是夜间模式，添加到帝瓦雷（Devialet）幻影，它允许你限制低音，让邻居高兴。

重塑您的类别主要需要更好地了解客户的深层生理需求，明智的定位，以及一定的胆识，才能使您打破密码并重新定义游戏规则。让我们希望这些热情洋溢的、敬业的设计师和成功的品牌能够激励您创建自己的独有风格。

This multisensory dimension of sound has been explored further via an unexpected but fine combination: "Devialet and Krug released a luxury box pairing 3 'lost' Ella Fitzgerald recordings (from her 1961 performance at the Concertgebouw in Amsterdam) with 3 'lost' editions of Krug Grande Cuvée," explains François, "and I liked very much the analogy, Krug CEO, Olivier Krug did between Champagne being a blend of different varieties of grapes and a music arrangement."
You can experience the Devialet sound in their stores in Paris, Frankfurt, or Shanghai Xintiandi, simply in the comfort of your own home, on Devialet Phantom, or wearing the recently launched Devialet Gemini true wireless earbuds. The Devialet area inside the Opéra Garnier in Paris is the most distinct with 18 Devialet Phantom loud speakers joining their harmonies to deliver an acoustic symphony. "Volume and depth are two different things" rounds up François and Devialet, that some call "the goosebumps company" is expanding rapidly but always trying to stay flexible and customer-friendly. Best example is the night mode, added to Devialet Phantom, that allows you to limit the bass and keep neighbors happy.
Reinventing your category requires mostly a better understanding of clients' deep physiological needs, a smart positioning, and a certain audacity that will enable you to break the codes and redefine the rules of the game. Let's hope these passionate and engaged designers and successful brands will inspire you to create your very own signature.

图 1.14 巴黎奥佩拉加涅尔剧院帝瓦雷音响区（经 Devialet 许可印刷）
Figure 1.14 Devialet sound area at Palais Garnier, Opéra de Paris, Paris
(printed with Devialet permission)

1.7 练习 Exercises

1A Y-3 品牌代码

阅读 1.3 中 Y-3 的故事，访问品牌网站：www.y-3.com 并回答下列问题：

问题 1: 运用本章学习的框架描述 Y-3 品牌代码。您可以在 DervalResearch 代戈微信上获得品牌代码模板：扫描二维码并键入关键词 **设计 1**。

问题 2:Y-3 品牌中哪些代码来自山本耀司（Yohji Yamamoto）？哪些来自阿迪达斯（Adidas）？请举例。

完成了？请核对第 21 页的参考答案 。

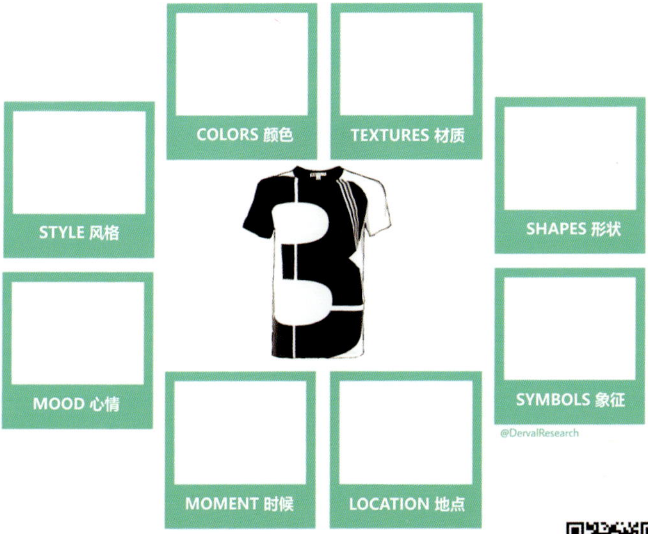

图 1.15 品牌代码模板（经由 DervalResearch 代戈提供）
Figure 1.15 Brand Codes Template (courtesy of DervalResearch 代戈)

1A Y-3 brand codes

Read the Y-3 story in section 1.3, visit the brand website www.y-3.com, and answer the following questions:

Question 1: Describe Y-3 brand codes using the framework learned in this chapter. You can get the Brand Codes framework on DervalResearch 代戈 wechat: scan the QRcode and type the keyword design1.

Question 2: Which codes come from Yohji Yamamoto? Which codes come from Adidas? Give examples.

Done? See the solution page 21.

1B 新产品或新类别

参观你喜欢的品牌店铺并描述其品牌代码。品牌能添加什么产品或品类来丰富产品组合？根据已描述的品牌代码，请清晰地就产品应在颜色、材质、形状、香味、品味、时代、地点、风格、基调、标志方面传达的信息提出你的建议。

Derval 代尔瓦勒色彩测试结果

- 你看到少于 20 个颜色的细微差别：你错过了相当多的颜色细微差别，必须有一个不规则的颜色受体分布，一些不太好覆盖的区域。好消息是你可能擅长其他感官技能，比如运动或 3D，你可能会穿黑色、米色和蓝色的衣服。

- 你会看到 20 到 33 种颜色的细微差别：你错过了一些颜色的细微差别，必须有一个规则的颜色受体分布，只有一些不太好覆盖的区域。你喜欢不同的颜色，因为你可以欣赏它们。

- 你会看到 34 到 38 种颜色的细微差别：你只是错过了几个颜色的细微差别，而且一定有一个很好的颜色受体分布。你喜欢色彩的微妙之处，因为你可以欣赏它们。

- 你看到 39 种颜色：恭喜！你可以看到所有的颜色细微差别，必须有一个非常完整的颜色受体分布。如果你有太多的受体，你甚至可能会被黄色刺激，所以这种颜色在你的衣柜里找不到。

- 你会看到超过 39 种颜色的细微差别：测试中只有 39 种不同的颜色，为什么？事实上，你对对比度的反应比对颜色的反应更强烈，可能是计算了分离条纹而不是实际的颜色 (Derval, 2018)。

1B New product or category

Visit a store you like and describe the brand codes. What product or product category could the brand add in order to enrich its assortment? Based on the brand codes just described, provide clear recommendations on how the product should be in terms of colors, textures, shape, scent, taste, moment in time, location, style, mood, and symbols.

Derval Color Test Results

- You see less than 20 color nuances: You missed quite some color nuances and must have an irregular distribution of color receptors, with some less well covered areas. Good news is that you are probably skilled at other sensory dimensions like maybe movement or 3D. You are likely to wear colors like black, beige, and blue.
- You see between 20 and 33 color nuances: You missed some color nuances and must have a regular distribution of color receptors, with just some less well covered areas. You enjoy different colors as you can appreciate them.
- You see between 34 and 38 color nuances: You just missed a couple of color nuances and must have a good distribution of color receptors. You enjoy colors subtleties as you can appreciate them.
- You see 39 color nuances: Congratulations! You could see all the color nuances and must have a very complete distribution of color receptors. If you have too many receptors, you might even be irritated by yellow, so this color would be nowhere to be found in your wardrobe.
- You see more than 39 color nuances: There are only 39 different colors in the test, how come? In fact, you respond more to contrast than to colors and probably counted the separation stripes instead of the actual colors (Derval, 2018).

练习 1A 答案

Y–3 的颜色、形状和基调体现山本耀司（Yohji Yamamoto）的风格。

Y–3 的材质、标志和风格反映阿迪达斯（Adidas）的风格。

两个品牌反映的地点和时刻同 Y–3 一致。

Y–3 的颜色中和了山本耀司推崇的黑色品味和阿迪达斯五彩缤纷的风格 (Derval, 2016)。

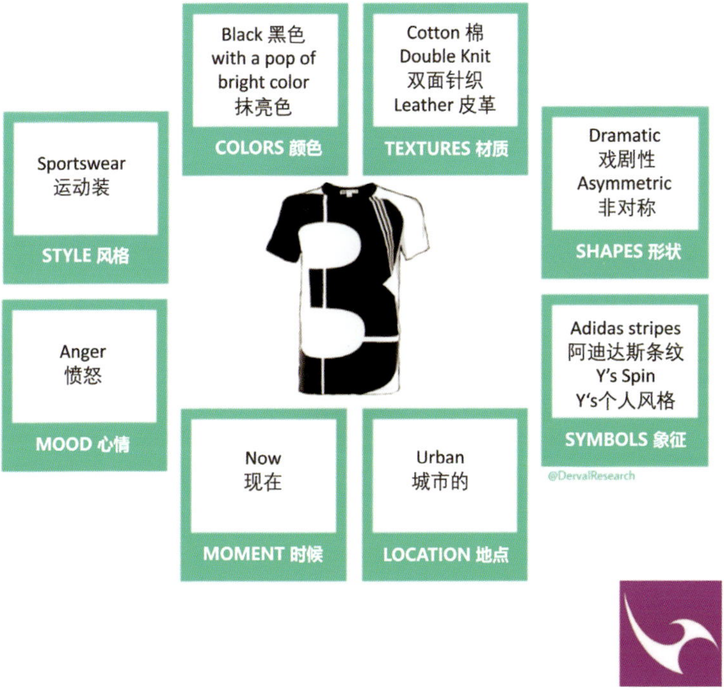

图 1.16 Y–3 品牌代码（经由 DervalResearch 代戈提供）
Figure 1.16 Y–3 Brand Codes (Courtesy of DervalResearch 代戈)

Solution to Exercise 1A

The colors, shapes, and mood are signed Yohji Yamamoto.
The textures, symbols, and style are signed Adidas.
The location and moment in time are common to both.
The colors are probably a compromise between Yohji's taste for black and Adidas' more colorful style (Derval, 2016).

1.8 要点 Take-aways

- 品牌代码必须鲜明
- 品牌代码应得到保护
- 记录品牌代码有助于揭示并保护品牌基因
- 品牌代码一旦确立之后，品牌能在忠于其身份的同时，更容易地对代码进行必要的更新和革新
- 如果一个产品类别已经饱和、被抛弃或被忽视，也许是时候重新设计它了！

- Brand codes have to be distinct
- Brand codes should be protectable
- Documenting brand codes helps reveal and protect the brand DNA
- Once the brand codes are clarified, it is much easier to update and revamp the brand whenever needed while staying true to the brand identity
- If a product category is saturated, abandoned, or ignored, maybe time has come to reinvent it!

1.9 术语表 Glossary

品牌代码

“当产品组合如施华洛世奇那样不断发展时，品牌代码对于在客户眼中保持一致和可信至关重要。”施华洛世奇执行董事会中国顾问、Bang&Olufsen 中国咨询委员会成员 Joan Ng 解释说，他们可以用一个词来概括自己的品牌代码：“闪耀”。(Derval,2016)

蓝海战略

蓝海战略是指创造和执行这样的战略举措，以释放无竞争的市场空间，使竞争对手变得无关紧要。(金和莫博涅 ,2005)

当 PlayStation 和 Xbox 为了谁的颜色最黑，谁的清晰度最高，谁的速度最快而打得不相上下时，任天堂出品了一个带有异乎寻常的触觉界面的白色小东西。通过转化非游戏玩家，这家日本品牌神奇地在这场销量巡回赛中问鼎冠军，Wii 的最终销售量超过了 1 亿台，比 PlayStation 和 Xbox 加在一起的总和还多。(Derval, 2022)

轻奢

凯特・丝蓓（Kate Spade）介绍了轻奢的概念：“她早期的手袋定价在 300 美元左右，这使得年轻的职业女性可以负担得起，她们想要的不仅仅是便宜的产品，但却买不起更传统的高端设计师产品。”(Ketchel,2018)

Brand codes

“When the product portfolio is evolving as in the case of Swarovski, brand codes are critical to stay consistent and credible in the eyes of the customers.” Joan Ng, China Advisor to the Executive Board at Swarovski, and Member of the China Advisory Committee at Bang & Olufsen, explained that they can summarize their brand codes in one word: “sparkle”. (Derval, 2016)

Blue Ocean Strategy

Blue Ocean Strategy is about creating and executing such strategic moves that unlock uncontested market space which renders competitors irrelevant. (Kim & Mauborgne, 2005)

While Playstation and Xbox were fighting to know which one was the blackest, the highest-definitioned, the fastest, Nintendo showed up with a white thingie with an unusual haptic interface. By converting even non-players, the Japanese brand managed the tour de force to outsell Playstation and Xbox consoles combined, and over 100 million units in Wii lifetime. (Derval, 2022)

Affordable luxury

Kate Spade introduced the notion of affordable luxury: “her early handbags were priced around the US$300 mark, making them affordable for young professional women who wanted more than the cheaper offerings but couldn’t afford the more traditional, high end designers.” (Ketchel, 2018)

1.10 参考文献 References

Derval, D. (2022). The Right Sensory Mix: Decoding Customers' Behavior and Preferences. Springer Nature, ISBN: 978-3-662-63794-4.

Derval, D. (2018). Designing Luxury Brands: The Science of Pleasing Customers' Senses. Springer Business & Management, ISBN: 978-3-319-71555-1.

戴安娜 · 代尔瓦勒 (Diana Derval). 奢侈品品牌营销：创建 · 实施 · 案例 (Luxury Brand Marketing). 东华大学出版社 (2016), ISBN: 9787566910424.

Derval, D., Bremer, J. (2012). Hormones, Talents, and Career: Unlock your Hormonal Quotient. SpringerBriefs, ISBN: 978-3-642-25712-4.

Derval, D. (2011). Réussir son étude de marché en 5 jours. Eyrolles, ISBN: 978-2-212-54958-4.

DervalResearch (2020). The science of colors: From product to experience design. Iversity, Springer Nature Group. https://iversity.org/en/courses/the-science-of-colors

Kim, W. C., & Mauborgne, R. (2005). Value innovation: a leap into the blue ocean. Journal of business strategy.

Grimmer, L., & Mortimer, G. (2018, June 5). Kate Spade, the archetypal new yorker, sold whimsical, affordable luxury to women. The Conversation. https://theconversation.com/kate-spade-the-archetypal-new-yorker-sold-whimsical-affordable-luxury-to-women-97854

第二章 Chapter 2

靶向感官剖面 Targeting a Sensory Profile

天衣无缝 Seamless

从时尚到香水，感官感知是一个关键的设计元素。在本章中，我们将了解目标客户的感官剖面如何帮助设计成功的产品，如丝芙兰 (Sephora)、黑袜子 (Blacksocks) 和马克西姆巴黎(Maxim's de Paris)所示。米其林明星厨师科拉多·米开拉佐（Corrado Michelazzo）和时尚艺术家埃德加·阿蒂斯（Edgar Artis）为我们带来了他们创作过程中的一个潜移默化的高峰。

目录

图 2.1 方方为 DervalResearch 代戈绘制的插画
Figure 2.1 Illustration by FangFang for DervalResearch 代戈

From fashion to fragrances, sensory perception is a key design element. In this chapter, we see how contouring the sensory profile of the target customers will help design winning creations, as demonstrated by Sephora, Blacksocks, and Maxim's de Paris. Michelin star Chef Corrado Michelazzo and fashion artist Edgar Artis give us a sneak peek into their creative process.

Content of this chapter

2.1 探索你的感官剖面：代尔瓦勒气味金字塔
Discover your sensory profile: the Derval Pyramid of Scents

香味营销可以真正地决定或破坏消费者的购物体验。大多数人能处理空气、食物和伴侣的气味，40% 的人会被花香、酒精和化学物质的气味刺激。如果你是一个零售商，计划推广合成香料（通常含有酒精和化学物质），或者整天烧香，那就很好了：你可能最终会让你想要吸引的消费者望而却步。

刺激性气味可能会引发哮喘、恶心和头痛，顾客的本能反应是尽快逃跑。他们常常不知道什么是他们不舒服的原因，因为他们太忙于结束它。顾客在商店的时间对销售有明显的影响。感官体验必须使他 / 她感到足够舒适，以便能够留下来完成交易。一般来说，没有气味总比错误的气味更好。

根据消费者的感官感知对消费者进行细分比按收入或年龄分组更有意义，并使品牌能够为每个目标群体和当地市场设计正确的产品和服务。

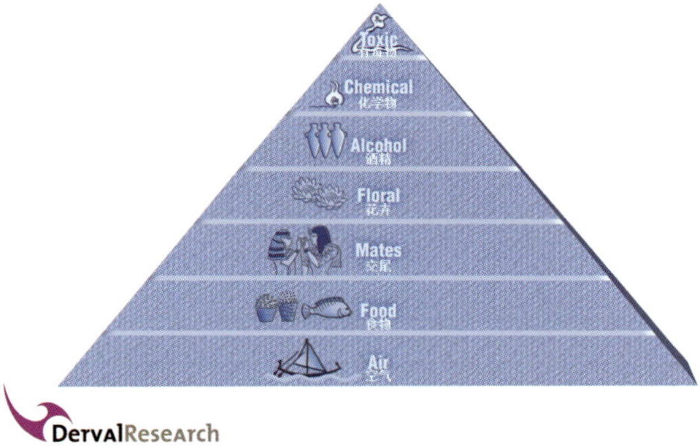

图 2.2 代尔瓦勒气味金字塔（图片来源：DervalResearch 代戈）
Figure 2.2 Derval Pyramid of Scents (courtesy of DervalResearch 代戈)

Scent marketing can truly make or break a consumer shopping experience. If most people can deal with the scent of air, food, and mates, 40% of the people are irritated by the scent of flowers, alcohol, and chemicals. Good to know if you are a retailer planning to spread synthetic fragrances (often containing both alcohol and chemicals), or to burn incense throughout the day: you might end up repulsing the consumers you wanted to attract.

An irritating scent might trigger asthma, nausea, and headache, and the customers' instinctive reaction is to run away as fast as they can. Often, they do not realize what is the cause of their discomfort as they are too busy putting an end to it. The time a customer spends in the shop has an obvious impact on sales. The sensory experience has to make him/her feel comfortable enough to stay in order to close the deal. In general, no scent is better than a wrong scent.

Segmenting consumers based on their sensory perception makes a lot more sense than grouping them by revenue or age, and enables brands to design the right products and services for each target group and local market.

根据 DervalResearch 代戈的观察和测量，我们可以将个体分成三组，分别考虑其气味偏好（Derval，2022）：

– 超级吸入者对化学物质非常敏感，比起花香，他们更喜欢水果味，比如橘子味。

– 中级吸入者喜欢水果味和淡淡的花香，如玫瑰。

– 非吸入者对化学物质不太敏感，喜欢辛辣或麝香的味道，以及薰衣草或茉莉花等强烈的花香。

例如，一个品牌可以为中级吸入者选择水果味的气味，而强烈的花香、麝香味甚至汽油味适合非吸入者，细分和定位是关键，因为没有所谓的“一种气味适合所有人”。

不同地区的人拥有不同的感觉处理能力。即使来自于同一个国家或者城市，人们的感觉处理能力也不同。为了帮助你们调整产品，DervalResearch 代戈在 25 个国家对上千名顾客进行了科学测试。以下测试的目的是为了确定你的感觉剖面。

问题 1：你偏爱 A. 浓缩咖啡或 B. 拿铁?

问题 2：你偏爱 A. 花香味或 B. 水果味香水？

问题 3：你偏爱 A. 流行音乐或 B. 其他音乐类型？

问题 4：你偏爱 A. 红色或 B. 蓝色？

问题 5：你偏爱 A. 棉或 B. 丝绸？

完成了？请参见第 39 页的解决方案。

Based on DervalResearch observations and measurements, we can split individuals into three groups regarding their scent preferences (Derval, 2022):
- Super-inhalers are very sensitive to chemicals and prefer fruit scents, like orange, to floral scents.
- Medium-inhalers enjoy fruit scents and subtle floral notes, such as rose.
- Non-inhalers are less sensitive to chemicals, and appreciate spicy or musky notes, strong floral scents like lavender or jasmin.

A brand could for instance select a fruity scent for medium-inhalers and a strong floral, a musky or even a gasoline scent for non-inhalers. Segmentation and targeting are key, as there is no such thing as “one-scent-fits-all”.

Sensory perception will vary between individuals. Huge differences can be measured depending on the country, province, and even in the same city. Here is a test to help identify your sensory profile. The test is based on scientific measurements conducted by DervalResearch 代戈 on thousands of people in over 25 countries.

Question 1: Are you more A. Espresso or B. Latte?

Question 2: Are you more A. Flower smell or B. Fruit smell?

Question 3: Are you more A. Pop music or B. Other types of music?

Question 4: Are you more A. Red or B. Blue?

Question 5: Are you more A. Cotton or B. Silk?

Done? See solution page 39.

2.2 丝芙兰的感官剖面 Sensory Profiles at Sephora

希西莉亚（Cécilia）今晚有约，下班后她去了巴黎丝芙兰（Sephora），在美容顾问的帮助下化上了精致的妆容，买了最爱的迪奥 Dior J'adore 香水。

艾薇（Ivy）今晚有约，下班后她去了上海丝芙兰（Sephora），在美容顾问的帮助下化上了精致的妆容，买了最爱的高田贤三 Flower by Kenzo 香水。

路易威登集团 LVMH 旗下的丝芙兰 Sephora 成立于 1970 年，在全球 30 个国家拥有 1700 多个销售点，并积极地在网上销售奢侈品品牌和自有品牌。可见要想品牌壮大并成功，美妆零售商也必须适应本土市场和本土消费者的感知。在法国，丝芙兰（Sephora）最畅销的是蕴含紫罗兰、玫瑰、西梅和香草的迪奥 Dior J'adore 香水，然而在中国最畅销的是蕴含山楂提取物、紫罗兰和香草的高田贤三 Flower by Kenzo 香水。由此可见， 希西莉亚（Cécilia）代表了典型的法国角色，而艾薇（Ivy）则代表了典型的中国角色。这两款香水都蕴含紫罗兰、香草和麝香，但是高田贤三 Flower by Kenzo 的香调比迪奥 J'adore 细腻。感觉知觉的变化是基于生理上的差异，这些差异很容易测量和比较人物角色之间的差异。对丝芙兰（Sephora）来说，了解客户的感官剖面对于选择合适的产品组合、实施合适的视觉营销和设计合适的宣传活动是非常重要的 (Derval,2016)。

比较希西莉亚（Cécilia）和艾薇（Ivy）感官知觉的最好方法是描述她们的感觉剖面。根据已知的产品进行描述分析，确定一个给定的产品——在这里香水，但它可以是一辆豪华汽车、游艇、米其林明星晚宴，或时装——对购买产品的用户来说，关键是找到描述他们感觉处理能力的 6 个标准关键，然后观察其中哪个标准对每个角色来说最重要。

Cécilia has a date tonight after the office and runs to Sephora in Paris for an ultimate beauty consultation and to refill her favorite perfume: Dior J'adore.

Ivy has a date tonight after the office and runs to Sephora in Shanghai for an ultimate beauty consultation and to refill her favorite perfume: Flower by Kenzo.

Sephora – part of the LVMH group – was founded in 1970 and counts over 1700 points of sale across 30 countries and is very active online, selling both luxury brands and its own label. The beauty retailer had to adapt to local markets and perception in order to grow and succeed. At Sephora, Dior J'adore is the best-selling perfume in France with notes of violet, rose, prune, and vanilla; and Flower by Kenzo is the best-selling perfume in China with notes of hawthorn, violet, and vanilla. So Cécilia would be a typical French persona and Ivy a typical Chinese persona. Both perfumes have violet and vanilla as well as musk in common but Flower is more subtle than J'adore. Variations in sensory perception are based on physiological differences that are easy to measure and to compare between personas. It is important for Sephora to understand the sensory profile of their customers to be able to select the right assortment, implement the right visual merchandising, and design the right communication campaigns (Derval, 2016).

The best way to compare Cécilia's and Ivy's sensory perception is to describe their sensory profiles. The idea is to identify for a given product – here a perfume but it could be a luxury car, a yacht, a Michelin star dinner, or a fashionable dress – the 6 criteria that are critical for the persona when purchasing and see which of these criteria are the most important for each persona.

感觉剖面的目的是为了检验产品是否符合用户的需求。我们可以用同样的方法描述用户的嗅觉阀值和产品的感官属性。这里的关键是从用户的标准处着手而非列举产品特点，因为有些特点可能不对消费者的购买产生影响。

我们可以使用这个极坐标图（图 2.3）来进行产品匹配。首先，我们需要明确 6 个既能区分不同用户又能决定他们是否购买的标准。对希西莉亚（Cécilia）而言，香调的持久度、麝香香调和花香香调是关键；但对于艾薇（Ivy）来说，重要的是花香、幽微，是否有化学成分也是她的主要考虑因素。我们按 1~10 级评估每一个标准对每一种用户的重要程度。1 代表毫不重要，10 代表非常重要。然后，我们分析其他产品，评估高田贤三 Flower by Kenzo 香水在每个标准下的得分，甚至可以直接将香水放进极坐标系。结果显而易见，高田贤三 Flower by Kenzo 正好适合艾薇（Ivy）。

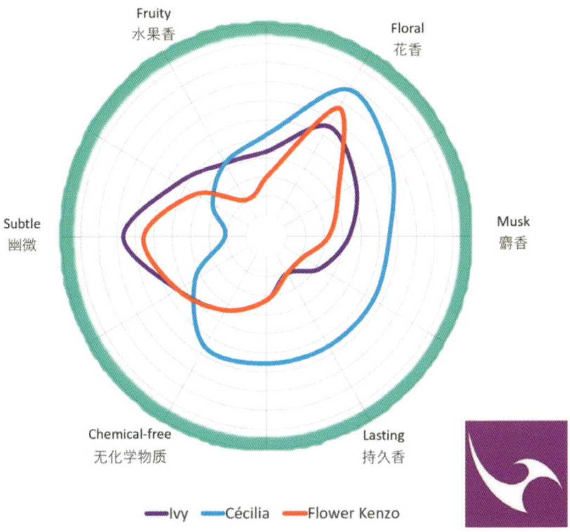

图 2.3 丝芙兰高田贤三感觉剖面（来源：DervalResearch 代戈）
Figure 2.3 Sephora and Kenzo Sensory Profiles (courtesy of DervalResearch 代戈)

The purpose of the sensory profiles is also to check if a product matches a persona. In the same way we described the threshold for each persona, we can describe the threshold for each product. The trick is to start with the persona's criteria instead of listing the product's features, as some of them might not influence the purchase at all.

We can use this polar graph (Figure 2.3) to do the product match making. First, we need to identify around 6 criteria that define and differentiate our personas and are critical in their purchasing decision. For Cécilia, lasting, musk, and floral are key. For Ivy floral is important as is subtle, and chemical-free is not a big concern. We evaluate for each persona how critical each criterion is on a scale from 1 to 10 for instance, with 1 being not important at all, and 10 being very important. Then we can do the same exercise for the products and evaluate how Flower by Kenzo rates on each of these criteria and even put the perfume on the polar graph. It then becomes obvious that Ivy and Flower by Kenzo are a good match！

2.3 黑袜子：右脚
Blacksocks: the right feet

强大的品牌故事也可以从讲述一个难忘的趣闻开始。

黑袜子（Blacksocks）商业帝国的建立源于公司的总经理萨米·李希尔一次颇为尴尬的经历。当时，他还是一名年轻的经理，作为顾问，他需要穿着精致的西装——当然是条纹的——出席一场相当重要的商务晚宴。可当他抵达晚宴现场时却发现那是一家日本餐厅，客人在用餐前需要脱鞋子。当他意识到自己一只袜子上有洞而且两只袜子一只是深灰色而另一只是黑色的事实后，觉得自己窘迫极了。这之后，他立马进行了市场调查，结果发现大多数男性不愿在买袜子上浪费时间。此外，商店里的袜子组合经常变化，所以就算你找到了一双理想的袜子，也不见得能一直买到。还有，恐怕没人想遭遇袜子有洞的尴尬吧(Derval, 2016)。

于是萨米·李希尔决定创造一个品牌，去解决这个关乎不少人的问题。

图 2.4 黑袜子 Blacksocks 制袜大师（图片来源：Blacksocks, 代尔瓦勒 2016）
Figure 2.4 Blacksocks Sockmaster (Photo courtesy of Blacksocks, Derval 2016)

A memorable anecdote can also be a very good starting point for a strong brand story.
Samy Liechti, the general manager of Blacksocks.com, built an empire upon a rather embarrassing anecdote. While he was still a young manager, he had to attend a very important business dinner. He put on his nicest suit – with stripes, of course – as he was a consultant. When he arrived, he discovered that the restaurant was Japanese and they had to take off their shoes. This was a big moment of embarrassment when he realized that one of his socks had a hole and that actually the other one was more dark gray than black. Doing a quick market research, he found out that most men do not want to waste time buying socks. Also, the assortment in shops rotates often. So even if you find the ideal pair of socks, soon it will be replaced. And nobody wants to be embarrassed with a hole in a sock (Derval, 2016).
So Samy Liechti realized that the creation of a brand had to solve a problem concerning several people.

因此基于这个商机，他设计了每月邮寄三到四双袜子的订袜方案——有点像订奶或古老的订报服务。萨米这么做不只是在卖袜子，而是创造了一项奢华的服务，因此到目前为止还没有一个竞争对手能和他抗衡。跟袜子一起寄出的不是发票，而是一封包含如何穿戴，如何搭配袜子和西装，如何挑选领带等礼仪信息的信。这些信息对他的目标客户群体非常有用。顾客能随时享受他的服务，无论什么问题都会受到高度重视，所以 30% 的新顾客由老顾客推荐。此外，25% 的顾客成为了品牌代言人。

黑袜子（Blacksocks）还借助最新的技术开发了令人兴奋和疯狂的新产品，如“智能袜”。内含 RFID 芯片的智能袜能让你在苹果手机 (iPhone) 应用上检查袜子从洗衣机取出后是否出现张冠李戴和遗漏的情况。“黑度表”应用则是另一种创新，有了它，你能在苹果手机 iPhone 上检查袜子的黑度，避免出现像萨米那样在日本餐厅中发现袜子一只深灰一只浅黑的尴尬。

随着该公司在国际上的扩张——美国代表了他们的第二大市场——在购买行为上的差异非常突出：

• 德国人倾向于更厚的秘鲁棉袜。

• 法国人倾向于选择更薄的丝光棉袜，也称为“fil d’écosse”。

最喜欢的质地似乎更依赖于服装习惯，而不是天气或季节性。喜欢厚袜子的男人穿运动鞋，而喜欢薄袜子的男人则穿商务鞋。感觉知觉呢？秘鲁的棉花比丝光的光滑。有趣的是，德国人对脚敏感的历史由来已久，他们鼓励开发像 Birkenstock 凉鞋这样的对脚友好的主要品牌。

So based on this opportunity, he designed the Sockscription, in which every month you receive by mail three to four pairs of socks – a little bit like you would receive your milk or in ancient times the newspaper. So far, no competitor can measure up with Samy because he is not only sending socks but he created a luxury service. Instead of joining an invoice to the socks, he is sending a letter with useful advice on etiquette, and how to match your socks with your suit and what types of ties to wear. All very useful information for his target business customers. The customer service is always available and any question is considered of the highest importance. Thirty percent of the new customers are recommended by existing customers. And 25% of the customers became brand ambassadors.

Using the latest technology, Blacksocks was able to release some exciting and crazy offerings like the “smart socks”. Thanks to an RFID chip in the socks, you can check on an iPhone app at the exit of the washing machine whether your socks are properly paired or not, and if some are missing. Another innovation is the “blackometer”. You can point your iPhone at your socks to identify how black they are and never end up again in the Japanese restaurant situation with a dark gray and a light black sock.

As the firm was expanding internationally – the U.S. represent their second market – they noticed differences in purchasing behavior:

• Germans prefer thicker Peruvian cotton socks.

• French prefer thinner Mercerised cotton socks, also called “fil d’écosse”.

The favorite texture is more dependent on clothing habits than on weather or seasonality. Men preferring thicker socks are wearing sneakers while men preferring thinner socks are wearing business shoes. What about sensory perception? The Peruvian cotton is smoother than the Mercerised one. Interestingly, Germans have a long history of foot sensitivity, encouraging the development of major foot-friendly brands like Birkenstock sandals.

2.4 与星共餐：米其林红色宝典
Dining with the stars: the Michelin Guide

每年米其林都会派神秘食客到餐厅，根据餐厅的食物质量和用餐环境，为主厨授予最高米其林 3 星的荣誉以及一颗新引入的绿色星级，以表扬可持续性 (Derval,2016)。美好的地方更在于为你指路帮你换轮胎以及为你提供食物的可都是同一个米其林先生噢!

米其林主打生活方式而不是产品种类。秉承着“一起前行”的定位，这家法国公司有价出售汇集了轮胎、地图、观光指南、导航工具和知名餐厅指南等信息的米其林指南。说到轮胎和食品，这些都是与安全相关的产品，可见米其林基于顾客对其产品品质的感知塑造了品牌的整体形象。虽然许多品牌企图超越米其林红色宝典,但可以肯定的是成功者寥寥。原因在于米其林不以产品技术，而以生活方式为基础的愿景，品牌因此能活跃在不同的市场中，同时还能保持消费者对品牌的清晰认识。

图 2.5 方方为 DervalResearch 代戈绘制的插画
Figure 2.5 Illustration by FangFang for DervalResearch 代戈

Every year secret tasters visit restaurants and judge the quality of the food and dining environment granting from zero to three coveted Michelin stars to the Chef, as well as a newly introduced green star to praise sustainability. Fascinatingly, it's the same Michelin Man who helps you find your way, change your tire, and serves you food!

Michelin opted for lifestyle instead of product category. With the positioning “moving forward together”, the French firm commercializes worldwide tires, maps, sightseeing guides, navigation tools, and the famous fine-dining Guide Michelin. In the case of safety-related products like tires and food, the whole brand image has been built on the perception customers have of Michelin's quality. What is sure is that many brands tried to outsmart the Michelin Guide but with little success. A vision based on a lifestyle rather than on a product expertise enables Michelin to play on very different markets without confusing customers.

为了更好地了解奢华餐厅和精致美食，我们采访了科拉多·米凯拉佐主厨(Derval,2016)。科拉多主厨目前是盖亚集团旗下的伊索拉意大利精品餐厅和位于香港国际金融中心的维多利亚港欧伯乐酒店的行政总厨。科拉多主厨和特罗斯葛罗和马克·维拉等知名主厨一起学烹饪。他的厨房在意大利开业两年后，他很快就获得了米其林的第一颗星。随之而来我肩负了更多责任："应该说这是一个激动人心但也是非常艰难的时刻，因为我同时要管理 6 至 7 个项目。那时候我瘦得太多了！"现在他是行政总厨，他有更多的时间专注于菜单。科拉多大厨在美食这样一个竞争激烈的领域取得成功的秘诀是，通过类比创造出令人惊叹的菜肴："我从日常生活活动中获得灵感：一杯莫吉托可以让我突然想到一种带有薄荷味的新鱼粉。"

To learn more about luxury restaurants and fine dining, we met with Michelin Star Chef Corrado Michelazzo (Derval, 2016). Chef Corrado is currently Executive Chef at Isola Italian fine dining restaurant, part of the Gaia Group, and overlooking Victoria Harbour in IFC Mall HK. Chef Corrado did his cooking classes with famous chefs like Troisgrois and Marc Veyrat and quickly got his first Michelin star only two years after opening his kitchen in Italy. The star brought a lot of additional responsibilities: "This was an exciting but also very tough time, as I was managing 6 or 7 projects simultaneously. I lost too much weight back then!" Now he is Executive Chef, he has more time to focus on the menu. Chef Corrado's secret ingredient to success in such a competitive field as gastronomy is to create stunning dishes by using analogy: "I get inspired by daily life activities: a Mojito can suddenly make me think of a new fish meal with a hint of mint."

2.5 埃德加·阿蒂斯（Edgar Artis）时装插画 Fashion Illustration with Edgar Artis

时装设计是一种多感官体验，其中光线、阴影，当然还有纹理起着重大的作用。时装设计师，Instagram 红人埃德加·阿蒂斯（Edgar Artis）分享了有关他惊艳的时装插画制作过程的一些秘密。埃德加·阿蒂斯（Edgar Artis）刚刚推出他的珠宝品牌 Sweven studios，他可以把任何东西都变成华丽的时尚插画：《苦涩的真相》插画是用咖啡制作的（图 2.6）。

图 2.6 埃德加·阿蒂斯的《苦涩的真相》（由埃德加·阿蒂斯提供）
Figure 2.6 Bitter Truth by Edgar Artis (courtesy of Edgar Artis)

Fashion design is a multisensory experience where light, shadow, and of course textures play a huge role. Edgar Artis, fashion designer and Instagram sensation, shares some secrets about his stunning fashion illustrations' making of. Edgar Artis – who is just launching his jewelry brand Sweven studios – can turn anything into a gorgeous fashion illustration: Bitter Truth illustration is made with coffee (Figure 2.6).

3 个步骤搞定时装插画：

步骤 1- 比例。我们一个接一个地画 9 个头和将它们分为几个部分，因为这有助于我们按比例绘制模特。Edgar 告诫道："我没有具体的步骤，因为我是一个自学成才的艺术家，我完全靠自己学会了如何画画，尝试并试验了多年，我找到了自己的方式去画一个人。我的手已经有了画脸部和身体的记忆。但我可以建议初学者开始使用被称为 9 个头的技术，基本上你用铅笔轻轻地画 9 个头，然后你把它分成身体的不同部分，你就有了时装插画的比例。我可以用图画更好地说明如何开始。"

步骤 2- 表达感情。"我从日常生活，周围所注意到的日常物品中获得灵感。我可以在街上或超市中看到一些物体，并产生创作插图的想法。我也从自己的感受，日常生活和体验中得到启发。" Edgar 透露："我的插图不仅基于以不同方式展示日常物品，而且还表达情感并将信息发送给看到它们的人。"

步骤 3- 完美心流。Edgar 透露了他最喜欢的工作环境："理想的时间是下午，因为那时候我非常专注并且光线对我的眼睛来说是完美的，但最重要的是我需要冷静、平和。当我紧张的时候，我不能画得很漂亮，结果并不完美，我不喜欢这个过程，我必须与自己和睦相处，我更喜欢独自一人画画，在这种情况下，我喜欢创作的过程。"

Nailing a fashion illustration in 3 steps:

Step 1 – Proportions. We draw nine heads on top of each other and divide them into parts, as it helps us draw a model proportionally. Edgar warns: "I don't have specific steps that I follow because I am a self-taught artist: I learnt how to draw all by myself, trying and experimenting for years. I found my own way of drawing a human; my hands already have the memory to draw the face and the body. But I can suggest for the beginners to start using the technique that is called 9 heads, basically you draw nine heads on top of each other – very lightly using the pencil – and then you separate it to different parts of the body and you have the proportions for fashion illustration."

Step 2. Express feelings. "I get my inspiration from daily life, daily objects that surround me and that I notice. I can see some object in the street or in a supermarket and get the idea of creating an illustration. I also get inspired from my own feelings; feelings that I go through in my daily life and experience." confides Edgar. "My illustrations are not only based on showing everyday objects in a different way but also expressing feelings and sending messages to people who see them."

Step 3 – The perfect flow. Edgar reveals his favorite working environment: "Ideally it's the afternoon because I am very focused that time and the light is perfect for my eyes, but most important is that I need to be calm, in peace. When I am stressed, I can't draw beautifully, the result is not perfect and I don't enjoy the process, I have to be in peace with myself and I prefer to be alone while I'm drawing, in that case I enjoy the process of creating."

2.6 豪华巧克力与马克西姆巴黎 Chocolate luxury with Maxim's de Paris

马克西姆巴黎（Maxim's de Paris）起源于巴黎，是一家具有传奇色彩的餐厅。著名作家、演员和大人物都曾在这间拥有“美好年代”即上世界 20 年代装饰风格的极具时尚气息的餐厅内用餐。令人震惊的是女演员碧姬·芭杜甚至有一次光着脚来到这里。该品牌也因其提供的餐饮服务而闻名于世，并在全球开了连锁店。例如，马克西姆巴黎（Maxim's de Paris）餐厅在北京经营其著名餐厅已有 30 多年的历史 (Derval,2016)。

由已故著名时装设计师皮尔·卡丹创立的马克西姆巴黎店，在输出概念时决定以巧克力为标志和核心产品。明确的核心产品对品牌而言始终非常有用，但这并不意味着品牌不能涉及包括香槟、小食、佐餐和甜品在内的其他美食。但以巧克力这类食品作为核心产品——有时热带国家推广很难成功，因为高温下会融化——倒给了马克西姆巴黎（Maxim's de Paris）一个优势。该品牌在零售领域和有礼品需求的 B2B 平台上非常活跃。

巧克力是真的很受欢迎，因为它有三个主要的让人上瘾成分：糖，咖啡因和可可碱 (Zoumas,Kreiser, & Martin,1980) 。有更多 μ（mu）阿片受体的人对蛋白质、可可碱和糖有反应。有更多 κ（kappa）阿片受体的人会喜欢薄荷或咖啡因。两种受体都太多的人注定凄惨 (Derval & Bremer,2012) 。所以我们喜欢的巧克力类型与我们的味蕾和阿片受体直接相关。不管怎么说，巧克力应该能最大限度地减少衰老带来的衰退，那么为什么要抵制呢？ (Moreira, Diógenes,de Mendonça,Lunet, & Barros,2016)

Maxim's de Paris is originally a restaurant in Paris. A legendary restaurant, where famous writers and actresses and powerful people used to have dinner in this very fashionable location with a "Belle Epoque" decoration. Once, actress Brigitte Bardot even shocked people by entering the restaurant with bare feet. The brand also became renowned for its catering and opened new locations around the world. In Beijing, for instance, Maxim's has been running its prestigious restaurant for over 30 years (Derval, 2016).

Founded by late Pierre Cardin, the famous fashion designer, Maxim's de Paris decided to use chocolate as its symbol and core product when exporting the concept. So having a clear core product is always very useful. It doesn't mean that the brand is not doing any other delicious things from champagne to snacks, savory and sweet. But having a strong core product like chocolate – sometimes difficult to manage in hot countries because it can melt – gave Maxim's an edge. The brand is very active in retail as well as in B2B with company gifting.

Chocolate is a true crowd pleaser, thanks to its three main addictive components: sugar, caffeine, and theobromine (Zoumas, Kreiser, & Martin, 1980). People who have more μ (mu) opioid receptors respond to proteins and to the theobromine and sugar. People who have more κ(kappa) opioid receptors will fall for mint or caffeine. People who have too many of both receptors are doomed (Derval & Bremer, 2012). So, the type of chocolate we prefer is directly linked to our taste buds and opioid receptors. Anyway, chocolate is supposed to minimize the decline faced with aging, so why resist? (Moreira, Diógenes, de Mendonça, Lunet, & Barros, 2016)

图 2.7 马克西姆巴黎巧克力 (图片来源 : 马克西姆巴黎 , Derval 2016)
Figure 2.7 Maxim's de Paris chocolates (photo courtesy of Maxim's de Paris, Derval 2016)

2.7 练习 Exercises

2A 黑袜感官剖面

阅读第 2.3 节中关于黑袜子的案例，观察针对法国市场和德国市场的袜子说明，并回答以下问题：

问题 1：确定同感官知觉相关并能影响潜在袜子购买者的 6 种标准。

问题 2：运用感觉剖面，标出 6 种标准，评估每个标准对法国 / 德国用户的重要性。您可以在 DervalResearch 代戈微信上获得感官剖面框架：扫描二维码并键入关键词 **设计 2**。

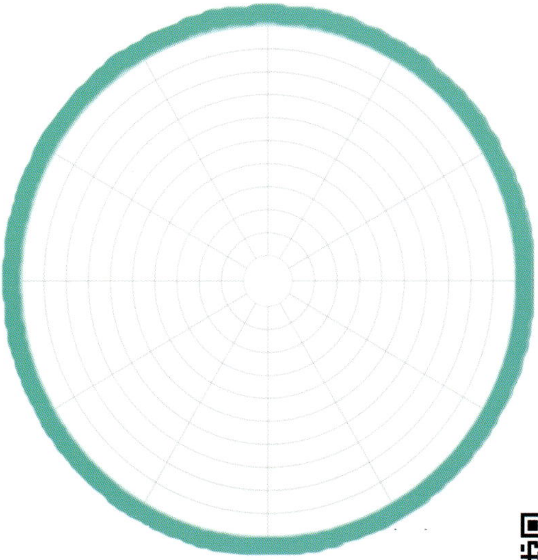

图 2.8 黑袜感官剖面（来源：DervalResearch 代戈）
Figure 2.8 Blacksocks Sensory Profiles (courtesy of DervalResearch 代戈)

2A Blacksocks Sensory Profiles

Read the case on Blacksocks in section 2.3 and check the Sockscription pieces intended for the German markets and those intended for the French market and answer the following questions:

Question 1: Identify 6 criteria linked to sensory perception that could influence the potential Sockscription buyers.

Question 2: Using the sensory profile template, put these 6 criteria as dimensions of the polar graph and evaluate how important each one is for a French persona and for a German persona. You can get the Sensory Profile framework on DervalResearch 代戈 wechat: scan the QRcode and type the keyword design2.

2B 艾薇的马克西姆巴黎 Maxim's de Paris

访问 Maxims-shop.com 商店或 Maxim’s 产品组合，准备一个礼盒，里面有 5 件非常适合艾薇角色的商品（见第 2.2 节）。证明您的选择是正确的。

感官剖面测试结果

问题 1

A：你对苦味不太敏感，而且任何食物你都能吃。

B：你对苦味挺敏感，避免吃太肥腻或甜的食物。

问题 2

A：你觉得大多数的香水和精油味道怡人。

B：你对酒精的味非常敏感，更喜欢清淡自然的香水。

问题 3

A：当你同他们交谈时有背影噪音，你需费力地聆听。

B：相比高音，你更喜欢低音。

问题 4

A：你能看清楚远处的物体，看红色和橙色时感到放松。

B：你能看清楚近处的物体，看蓝色时感到放松。

问题 5

A：你对某种纺织品和温度变化敏感。

B：任何面料你都喜欢，而且你更注重衣服的形状和图案。

2B Maxim's de Paris for Ivy

Visit Maxims-shop.com or Maxim’s assortment and prepare a gift box with 5 items that are perfect for the Ivy Persona (see section 2.2). Justify your selection.

Sensory Profiles Test Results

Question 1

A: You are not too sensitive to bitterness and you can eat almost anything.

B: You are quite sensitive to bitterness and avoid foods that are too fatty or sweet.

Question 2

A: You find most perfumes and essential oils pleasing.

B: You are rather sensitive to alcohol smells and prefer light and natural perfumes.

Question 3

A: You struggle to follow a conversation when there is background noise.

B: You prefer low sounds to high pitch sounds.

Question 4

A: You have a good vision from far away and are more relaxed with colors like red and orange.

B: You see better from close by, and blue relaxes you.

Question 5

A: You are irritated by certain types of textiles and are sensitive to temperature changes.

B: You enjoy every type of fabric and pay more attention to the shape and pattern of the clothing.

2.8 要点 Take-aways

- 香味营销可以真正地决定消费者的购物体验
- 感官剖面的目的是帮助检查产品是否符合人物角色
- 随着品牌在国际上的扩张，或是应对新类型的客户，他们会注意到感官剖面和购买行为的差异
- 设计是产品，也是体验
- 个人之间的感官差异如此之大，以至于品牌不能取悦所有人，而需要专注于他们优先考虑的角色
- 实现可持续发展的最佳途径是实施有针对性的产品开发

- Scent marketing can truly make or break a consumer shopping experience
- The purpose of the sensory profiles is to help check whether a product matches a persona
- As brands expand internationally – or tackle new types of customers – they notice differences in sensory profiles and purchasing behaviors
- Design is about products but also experiences
- Sensory variations among individuals are so huge that brands cannot please everybody and need to focus on their priority personas

■ The best way to achieve sustainability is to implement targeted product development

2.9 术语表 Glossary

神经科学

“神经科学的发展正在获得动力。神经科学已经积极介入人格本质、人格特征、人格特质遗传力的研究。神经科学即将改变我们的人性观念。”(Arakelov et al.,2010)

味蕾

“我们舌头上的味蕾数量可以在每平方厘米 11 到 1100 个之间变化。每个味蕾有助于感知苦、甜、盐、酸和鲜味相当于肉类、奶酪和蘑菇中的谷氨酸盐味道。” (Derval,2022)

感知

“研究表明，对自然环境的感知与人类健康之间存在着某种关系。我们的假设是，人们从某些维度来看待绿色空间，其中一些维度在帮助人们摆脱压力方面比其他维度更重要、更受欢迎。” (Grahn, & Stigsdotter,2010)

Neurosciences

“Development of neurosciences is gaining impetus. Neurosciences already actively interfere with the study of personality nature, its features, heritability of personality traits. Neurosciences are about to change our ideas of human nature.” (Arakelov et al., 2010)

Taste buds

“The number of taste buds we host on our tongue can vary between 11 and 1,100/cm^2. Each taste bud helps to perceive bitter, sweet, salt, sour, and umami (umami corresponds to the glutamate taste present in meat, cheese, and mushrooms).” (Derval, 2022)

Sensory perception

“Research indicates a relationship between sensory perception of natural environments and human health. Our hypothesis is that people perceive green spaces in terms of certain dimensions, where some dimensions are more important and preferred than others with respect to restoring people from stress." (Grahn, & Stigsdotter, 2010)

2.10 参考文献 References

Derval, D. (2022). The Right Sensory Mix: Decoding Customers' Behavior and Preferences. Springer Nature, ISBN: 978-3-662-63794-4.

Derval, D. (2018). Designing Luxury Brands: The Science of Pleasing Customers' Senses. Springer Business & Management, ISBN: 978-3-319-71555-1.

戴安娜·代尔瓦勒 (Diana Derval). 奢侈品品牌营销：创建· 实施·案例 (Luxury Brand Marketing). 东华大学出版社 (2016), ISBN: 9787566910424.

Derval, D., Bremer, J. (2012). Hormones, Talents, and Career: Unlock your Hormonal Quotient. SpringerBriefs, ISBN: 978-3-642-25712-4.

Derval, D. (2011). Réussir son étude de marché en 5 jours. Eyrolles, ISBN: 978-2-212-54958-4.

Zoumas, B. L., Kreiser, W. R., & Martin, R. (1980). Theobromine and caffeine content of chocolate products. Journal of Food Science, 45(2), 314-316.

Moreira, A., Diogenes, M. J., de Mendonca, A., Lunet, N., & Barros, H. (2016). Chocolate consumption is associated with a lower risk of cognitive decline. Journal of Alzheimer's disease, 53(1), 85-93.

Arakelov, G. G., & Glebov, V. V. (2010). Psychology future in development of neurosciences. Psychology in Russia: State of the Art, 3, 181-194.

Grahn, P., & Stigsdotter, U. K. (2010). The relation between perceived sensory dimensions of urban green space and stress restoration. Landscape and urban planning, 94(3-4), 264-27.

第三章 Chapter 3

提供独特体验 Delivering a Unique Experience

万紫千红 A riot of colors

设计就是提供一种独特的体验。客户偏好和品牌一致性是关键，我们将在本章宝马和高田贤三（Kenzo）案例中看到，而像角色这样的强大工具将有助于关注真正重要的东西。适应从中国到澳大利亚的每一个角色是一条必经之路，红牛就是一个很好的榜样。我们还将向盟可睐（Moncler）首席执行官 Remo Ruffini 学习如何改造品牌和提升客户体验。

目录

图 3.1 方方为 DervalResearch 代戈绘制的插图
Figure 3.1 Illustration by FangFang for DervalResearch 代戈

Design is about delivering a unique experience. Customer preferences and brand consistency are on the critical path, as we will see in this chapter with the BMW and Kenzo cases, and a powerful tool like persona will help focus on what really matters. Adapting to each persona, from China to Australia, is the way to go, and Red Bull is a great example to follow. We will also learn from Moncler CEO, Remo Ruffini, how to revamp a brand and boost the customer experience.

Content of this chapter

3.1 颜色偏好和视觉
Color Preferences and Vision

最近关于视觉错觉的帖子促使我分享了一些神经眼科的洞察。

看看这张图片：它是在动还是不在动?

结果参见第 57 页。

我第一次看到来自我的同事北冈教授的这个紫色和黄色的图像移动是在阿姆斯特丹。不久后我意识到它在其他城市也是移动的。

原因如下。也许你还依稀记得在物理课上打盹儿的时候，老师突然拿出一个玻璃棱镜，并在上面照射了光束，创造出一道漂亮的彩虹。

图 3.2 北冈明佳"月光照耀下的小海龟"（许可印刷）了解错觉详情，请访问京都立命馆大学网站上有关北冈明佳教授的网页：http://www.ritsumei.ac.jp/~akitaoka/index-e.html
Figure 3.2 "Baby turtles lightened by the moon" by Prof. Akiyoshi Kitaoka (printed with permission, visit Prof. Kitaoka's page on Ritsumeikan University of Kyoto's website for more illusions: http://www.ritsumei.ac.jp/~akitaoka/index-e.html)

Recent posts about optical illusions motivated me to share some neuro-ophthalmological insights.
Look at the picture: is it moving or not moving?
See the results page 57.
First time I saw this purple and yellow image – from my colleague Prof. Kitaoka – move, it was in Amsterdam. I soon realized it was also moving in other cities.
The reason is as follows. Maybe you vaguely remember while napping during the physics class, the teacher suddenly pulled out a glass prism and beamed light on it, creating a beautiful rainbow.

原因在于我们的眼睛。人们可见的光波长介于紫外线和红外线之间，即400至700纳米。每一种光线都有其特定的波长：紫色光接近400纳米，红色光接近700纳米(Derval,2022)。

彩虹之所以出现，是因为每一种颜色都有不同的波长（图3.3），并相应地从左到右折射到不同的点：

- 首先是紫色、蓝色和靛蓝，它们是短波
- 然后是绿色，它在中间有点不确定
- 最后是黄色，橙色和红色，它们是长波

类似地，当观看图片时，紫色将在视网膜的前部映入我们的眼睛，而黄色将在视网膜的后面映入我们的眼睛。这种现象被称为色差。我们的眼睛在两种颜色之间来回移动会产生运动的错觉（村上等人,2006）。

眼肌帮助眼睛改变焦点：拉紧和放松，从短波长颜色（紫色）切换到长波长颜色（黄色）。一些有强壮眼肌的人可以三下两下地来回移动，而其他人则会头晕目眩，甚至打个“特彩哈欠”（=呕吐）！

图3.3 以纳米为单位的颜色典型波长（经由DervalResearch代戈许可）
Figure 3.3 Colors' typical wavelength in nanometer (courtesy of DervalResearch 代戈)

Let's look into the human eye to explain why. People can see colors with a wavelength higher than ultra-violet and lower than microwave radiations, in the 400 to 700 nanometers (nm) range. Each color has its wavelength: purple close to 400 and red close to 700 nm (Derval, 2022).

The rainbow appears because each color has a different wavelength (Figure 3.3) and is refracted accordingly to a different spot from left to right:

- first purple, blue, and indigo, which are short wavelengths
- then green, which is a bit undecided in the middle
- and finally yellow, orange, and red, which are long wavelengths

Similarly, when looking at the picture (Figure 3.2), the purple will hit our eye at the front of the retina and the yellow at the back of the retina. This phenomenon is called chromatic aberration. Our eyes going back and forth between the two colors create the optical illusion of movement (Murakami et. al, 2006).

An ocular muscle helps the eye change focus: tensing and relaxing, switching from the short wavelength color (purple) to the long wavelength color (yellow). Some people with strong ocular muscles can go back and forth seamlessly, while others will get borderline dizzy or even do a "technicolor yawn" (= vomit)!

3.2 驾驶宝马的角色 Personas driving BMW

有些人渴望过“宝马香车”的生活。宝马（BMW）也确实是全球标志性的汽车品牌。宝马 BMW 全称为“Bayerische Motoren Werke”，意为德国巴伐利亚州制造的汽车。事实上品牌早在 1916 是制造飞机的，最初的名字是 Bayerische Flugzeug Werke，缩写为 BFW，即巴伐利亚制造的飞机马达。

宝马（BMW）总是在行业竞争中处于领先地位，为汽车注入创新的活力，令汽车配件兼具美观和实用性，同时不忽略品牌核心：性能。大多数宝马（BMW）7 系车配备的菲利尔（FLIR）夜视系统就是一个极好的例子，宝马（BMW）夜视系统采用了最新的红外线技术，动态光点功能能在黑暗中发现行人和动物并向他 / 它投射光线。此外，轻轻一踢，后备箱就能打开是另一个非常实用的功能，这是不是很棒呢！针对运动和操作开箱指令的镭射灵敏装置是这背后的秘密 (Derval,2016)。

图 3.4 方方为 DervalResearch 代戈绘制的插图
Figure 3.4 Illustration by FangFang for DervalResearch 代戈

Some people want to live the “baomaxiangche” lifestyle. BMW is indeed an emblematic car brand around the world. BMW stands for “Bayerische Motoren Werke” which means motors from the Bavarian province in Germany. Actually, the brand was originally called BFW for Bayerische Flugzeug Werke, which means Bavarian aircraft motors, as back in 1916 the firm was working on airplanes.

Always on top of the game, BMW fuels its cars with innovation and proposes a winning combination of aesthetical and practical accessories, not leaving aside its focus: performance. A great example is the night vision system provided by FLIR on most BMW series 7. Using the latest infrared technology, BMW Night Vision with Dynamic Light Spot detects people and animals in the dark and sheds light on them. Another very useful feature is to be able to open the trunk of the car just by kicking it with the foot: How awesome! The secret behind is a laser sensitive to the movement and operating the trunk opening (Derval, 2016).

让我们认识一下加里（图 3.5）——他是一位电商顾问和快乐的宝马车主。加里注重汽车更好更快的性能，因此他享受快节奏的电子商务工作环境并热衷科技。当他不在开车的时候——他选择了一份可以让他每天开车上班的顾问工作——他会在 Xbox 上玩赛车游戏，只要他老婆允许！会晤客户间隙，他喜欢边在露台咖啡馆品尝浓咖边浏览网上最新的技术配件。

他的荷尔蒙商趋于平衡，特征是注重地位，性格非常外向，驾车时有超速的嫌疑——详见第五章荷尔蒙商内容。他很有地位，很外向，有点不尊重速度限制。

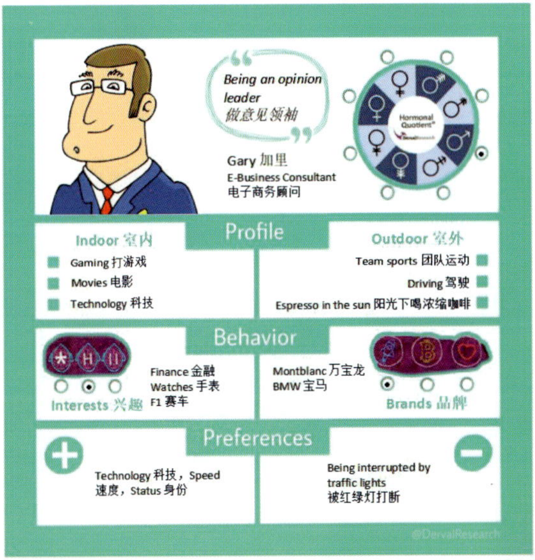

图 3.5 宝马 BMW 角色示例（框架经由 DervalResearch 代戈许可）
Figure 3.5 Example of BMW persona (Framework courtesy of DervalResearch 代戈)

Let's meet Gary (Figure 3.5) – an e-business consultant and happy owner of a BMW. Gary is all into performance, delivering the best, the fastest. That is why he enjoys the fast-paced e-business environment and is into technology. When not piloting his car – he picked a consulting job to be able to drive for work every day – he plays racing games on his Xbox, when his wife permits! In between appointments, he loves to check the latest technological gadgets online while enjoying a strong coffee at a café terrace.

His Hormonal Quotient® is balanced – we will see that in more detail in chapter 5. He is into status, very outgoing, and has a slight tendency not to respect the speed limits.

角色是在已有的顾客形象基础上补充用户清晰的动机、职业、爱好和生活方式的顾客代表，也是你能在人脉圈中和大街上辨认出的用户。

角色技巧被苹果（Apple）或飞利浦（Philips）等成功企业运用，这也将帮助我们明确目标顾客的形象、期望、兴趣和偏好。

为了描述角色，我们会记录以下问题：

■ 他们生活的动力是什么？
■ 他们年龄多大？收入一般是多少？
■ 我们了解他们的职业、爱好、教育水平吗？
■ 他们有家庭、孩子、宠物吗？
■ 别犹豫，只管用具体到与你的产品或项目相关的问题来完善它。

为了描述合理的角色，我们的想法是关注他们身上的共同点和事件。例如，如果有两位商务顾问名叫加里，角色就以加里命名。如果大多数顾客年龄在 34 岁左右，即使有的 25 岁，有的 48 岁，角色的年龄也定为 34 岁。这么做的关键是挑选顾客身上出现的高频率特征，避免在描述中出现平均数值和区间范围，从而获得清晰准确的模型。当你朋友的朋友看着这位用户模型说“我认识这家伙”时，你就成功了。

一般而言，一个品牌有 7~8 种角色，每一种都有其特定的感官知觉。

了解角色有助于开发正确的产品特征。

A persona is a representative customer, inspired by existing customers, with a clear motivation, job, hobbies, and lifestyle, someone that you can identify in your network, that you can recognize in the street.

The persona technique is used by successful companies, like Apple or Philips, and will help us identify the profile of target customers, their expectations, their interests, and their preferences.

To describe a persona, we would document the following questions:

■ What are their motivations in life?
■ What is their age and typical salary?
■ What do we know about their job? Hobbies? The type of studies they did?
■ Do they have a family? Children? Pets?
■ Do not hesitate to complete with questions more specific to your product or project.

To properly describe a persona, the idea is to focus on common points and occurrences. For instance, if two business consultants are called Gary, let's call him Gary. If most of them are around 34 years old, let's put 34, even if some are 25 years old and some 48. The point is to avoid averages and intervals and pick instead the most frequent features in order to get a clear and accurate picture. You know your persona is successful when people in your extended network look at it and say “I know this guy”.

Usually, a given brand has 7 to 8 Personas, and they each have specific sensory perception, needs and preferences. Knowing the persona helps with developing the right features.

3.3 红牛：金色还是蓝色？
Red Bull: Gold or Blue?

倘若找准了正确的目标客户，任何品牌都可以做到顶尖。

尽管口味怪异而且不在超市销售，红牛仍在全世界范围内取得了成功。

怎么做到的？通过完美标识出他们的目标客户和最能盈利的角色的方法。

这个品牌明确了 5 种购买罐装饮料的角色：派对迷、学生、司机、商务人士和运动员。在这 5 种角色里，极限运动的狂热爱好者是贡献利润最多的目标客户。

红牛现在专注于举办如 X Fighters 的国际性运动赛事。好主意，因为大多数红牛罐头都是在这些疯狂的自由式摩托车表演期间出售的！

关于红牛的一切都是令人兴奋的，包括罐的颜色——巧妙地适应当地。

DervalResearch 代戈最近在视觉和眼科研究协会（ARVO）年会上进行的一项研究表明，我们对颜色的偏好与我们的视觉有关。

图 3.6 方方为 DervalResearch 代戈绘制的插图
Figure 3.6 Illustration by FangFang for DervalResearch 代戈

Any brand can make it to the top provided they target the right customers.
In spite of a weird taste and no access to supermarkets, Red Bull managed to become a worldwide success.
How? By perfectly identifying their target customers and most profitable Personas.
The brand identified 5 Personas buying their cans: clubbers, students, drivers, business people, and sports people. Among the 5 Personas, the extreme sports aficionados revealed the most profitable target customers.
Red Bull now focuses on hosting international sports events like the X Fighters. Good idea as most of the Red Bull cans are sold during happenings like these crazy freestyle motocross shows!
Everything about Red Bull is exciting, including the color of the can – cleverly adapted locally.
A recent research conducted by DervalResearch 代戈 and presented at the annual meeting of the Association for Research on Vision and Ophthalmology (ARVO) shows that our preferences in colors are linked to our vision.

这是一个简单的物理学问题。

你还记得色差，即每种颜色在眼睛里的折射不同(图 3.7)。紫光照射在视网膜前部，红光照射在视网膜后部，绿光照射在两者之间。如果我们现在也考虑到角色的视觉敏锐度，它会变得很吸引人。

F 代表眼球中的焦点，位于视网膜内，也是所有颜色的光线透过水晶体汇聚之处，但是焦点的确切位置因人而异。

- 远视者的眼球较短，F 点在视网膜后部。长波长的颜色，比如黄色(也可以是金色)或红色会撞击视网膜的焦点，让人感觉放松。

- 近视的人，或近视眼聚焦光在视网膜前面。短波长的颜色，比如紫色或蓝色，会撞击视网膜的焦点，会被认为是令人放松的。红色会迫使他们的眼肌紧张，可能会兴奋一段时间，但也可能会变得刺眼 (Derval,2018)。

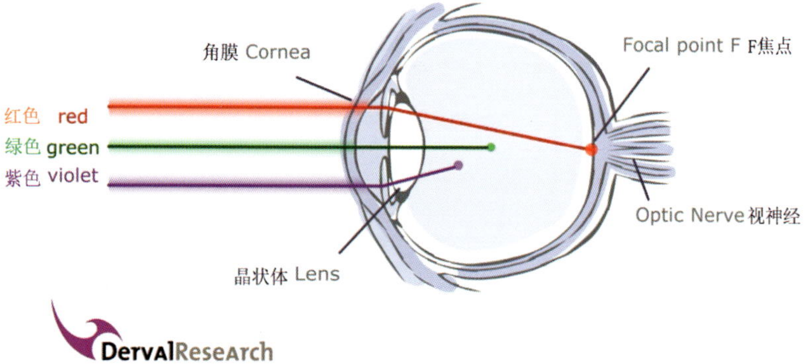

图 3.7 视觉和颜色偏好（经由 DervalResearch 代戈提供）
Figure 3.7 Vision and color preferences (courtesy of DervalResearch 代戈)

It's all a matter of simple physics.

You remember the chromatic aberration, where each color refracts differently on the eye (Figure 3.7). A violet light will hit the front of the retina, a red light the back of the retina, and a green light somewhere in between. If we now also take into account the persona's visual acuity, it gets fascinating.

F is the focal point of the eye, located in the retina, where all color waves meet after passing through the lens. The exact location of the focal point varies between individuals:

- Far-sighted people have a shorter eyeball and the F point is at the back the retina. A long wavelength color like yellow (gold also) or red will hit their retina at the focal point and will be perceived as relaxing.

- Near-sighted people, or myopes, focus light in front of the retina. A short wavelength color like purple or blue will hit their retina at the focal point and will be perceived as relaxing. Red will force them to tense their ocular muscle, and can be exciting for a while but might become irritating (Derval, 2018).

中国是世界上近视人数最多的国家，近视人数超过 4 亿人，占总人口的三分之一，占年轻人的三分之二。在过去的 30 年里，美国近视的比例也从 25% 上升到 41%。而绝大多数的澳大利亚人没有近视 (Derval,2022)。

在中国，近 70% 的新一代是近视，用红色包装礼物会让礼物看起来非常令人兴奋。而在只有 13% 的人是近视的澳大利亚，情况正好相反 : 蓝色是一种令人兴奋的颜色。

同样的颜色会有不同的影响，红牛完美地适应了当地的角色。

在中国，红牛是金色的，这是一种令人兴奋但又不太刺眼的颜色。在澳大利亚，红牛罐是蓝色的，这是一种令人兴奋的颜色。而在法国，你可以看到设计很复杂，因为人们甚至不能对国旗的颜色达成一致 (图 3.8)!

图 3.8 澳大利亚和法国国旗的颜色（经由 DervalResearch 代戈提供）
Figure 3.8 The colours of the Australia and France (courtesy of DervalResearch 代戈)

China has the biggest population of near-sighted people in the world with over 400 million individuals concerned, which represents one third of its population but two thirds of the youth. In the past 30 years, the proportion of near-sighted people in the USA increased also from 25% to 41%. Whereas the vast majority of Australians, for instance, are not near-sighted (Derval, 2022).

In China where almost 70% of the new generation is nearsighted, wrapping gifts in red will make the present look very exciting. In Australia, where only 13% of the people are nearsighted it is the exact opposite: blue is an exciting color.

The same color will have a different impact and Red Bull perfectly adapted to local personas.

In China, the Red Bull can is gold, which is an exciting but not too irritating color. In Australia, the Red Bull can is blue, which is an exciting color. And in France, you can see it is complicated to do design as people could not even agree on the color of their flag (Figure 3.8)!

3.4 统一是关键：夏之恋花样年华淡香水（Flower by Kenzo ）Consistency is key: Flower by Kenzo

艾薇（Ivy）不仅喜欢夏之恋花样年华淡香水（你记得她在第二章里出现过），她还喜欢它的包装。

我们对艾薇（Ivy）进行了检查，发现她不是近视眼患者，所以高田贤三 Flower by Kenzo 香水的中性白色背景和红色点缀令她眼睛放松。这明确了观察中国消费群体而不是依靠“平均”数据的重要性。高田贤三 Flower by Kenzo 非常适合在中国市场的丝芙兰（Sephora）销售。这也表明了奢侈品零售商的产品组合适应本地情况的重要性。

高田贤三（Kenzo）正是运用正确的香调（花香）和正确的包装（红色）才吸引了艾薇。

无论品牌进入哪个城市、省或大洲，本土化调整是业务拓展的关键。重要的是品牌能先明白它在发源地成功的原因，然后再向其他地区输出品牌概念。不同城市的人各不相同，即使在同一城市，从事不同工作，拥有不同爱好的人的感知情况也各不相同。感觉剖面丰富的角色将有助于比较人物角色的感知——就像我们在高田贤三（Kenzo）对迪奥（Dior）案例中所做的那样——并识别与现有产品的差距，促进与全球品牌的匹配。

Ivy not only loves the smell of Flower by Kenzo (you remember her from chapter 2) but she also likes the packaging.

We checked and Ivy is not nearsighted so that Flower by Kenzo is relaxing, with a touch of red and a neutral white background. This confirms the importance of observing the target consumer groups instead of relying on “average” statistics. So, Flower by Kenzo is the perfect product for Sephora China. This shows the importance for a retailer selling luxury products to adapt the assortment locally.

Kenzo did well by combining the right smell (floral) with the right packaging (red) to seduce Ivy.

图 3.9 方方为 DervalResearch 代戈绘制的插图
Figure 3.9 Illustration by FangFang for DervalResearch 代戈

Adapting locally is key for expanding a business. Whether a brand changes city, province, country, or continent, it is important to first understand the reasons behind its success in its hometown before exporting the concept. People are different from one city to another and even within the same city, depending on their job or hobbies. The persona enriched by the sensory profiles will help compare personas’ perceptions – as we did in the Kenzo vs Dior case – and identify the gaps with existing products and facilitate match-making for global brands.

3.5 与盟可睐（Moncler）首席执行官雷莫·鲁菲尼一起重塑时尚品牌 Revamping a fashion brand with Moncler CEO, Remo Ruffini

成功的设计总是针对特定的角色。在一次独家采访中，Moncler 设计师兼所有者 Remo Ruffini 解释了丰富多彩和大胆的设计如何帮助他重塑品牌，该品牌现已成为全球前 25 名奢侈品牌之一 (Derval,2016) 。

关键 1 - 伟大的产品和卓越的历史

盟可睐（Moncler），一个由两个登山者在 1952 年创立的品牌，以为奥林匹克队配备绗缝夹克而闻名。Remo Ruffini 以创意总监的身份加入了这个日渐衰落的品牌，并于 2003 年将其买回，随后于 2013 年首次公开募股（IPO）将其引入股市。他引入了更加丰富多彩的设计，激烈的视觉营销，将恐龙或鲨鱼放在场景中，并在迈阿密 (图 3.10) 或香榭丽舍 (图 3.11) 等战略要地建立了强大的分销渠道。我们有幸采访了 Remo Ruffini，并在讨论成功的奢华时装设计的关键要素时，Remo 回顾了他投资 Moncler 的动机："当我在 2003 年决定购买 Moncler 时，这个品牌的伟大传统和历史绝对是独一无二的，这是我想重新点燃的第一件事。

图 3.10 迈阿密巴尔港的 Moncler 窗口（经由 Moncler 许可）
Figure 3.10 Moncler Window in Miami Bal Harbour (courtesy of Moncler)

A successful design is always targeted for a specific persona. In an exclusive interview Moncler designer and owner Remo Ruffini explains how colorful and daring designs helped him revamp the brand, now part of the top 25 luxury brands worldwide (Derval, 2016).

Key 1 - A great product and an outstanding history

Moncler, a brand created by two mountaineers in 1952, became famous for equipping the Olympic Team with quilted jackets. Remo Ruffini joined the declining brand as a creative director and bought it back in 2003, before introducing it on the stock market with an IPO in 2013. He introduced more colorful designs, fierce visual merchandising—putting in scene dinosaurs or sharks—and built strong distribution channels in strategic places such as Miami (Figure 3.10) or the Champs-Elysées (Figure 3.11). We had the chance to interview Remo Ruffini and when discussing the key ingredients for a successful luxury fashion design, Remo comes back on the motivations behind him investing in Moncler: "When I decided to buy Moncler in 2003, the great heritage and history of the brand was absolutely unique, and that was the first thing I wanted to reignite.

很难找到一个有着深厚传统渊源的品牌。目标是从品牌的起源开始，将其带入未来。我们的目标是在全世界推行‘全球羽绒服’战略。很重要的一点是要告诉大家，Moncler 是所有场合的夹克。品牌的演变已成为城市永不放弃的运动精神，一直存在于 Moncler 的灵魂。这是我们真正想要的，从一个伟大的产品和杰出的历史开始，这是一个奇妙的‘冒险’。无需存档，但绝对需要开发。”(Derval,2016)

关键 2 - 盟可睐（Moncler）大胆展示橱窗

盟可睐（Moncler）刚刚在巴黎的主要购物街香榭丽舍大道开了一家新店，通过令人惊叹的视觉营销和寻宝活动让路人和媒体目瞪口呆，让这款服装出现在了前所未有的场景中。当谈到成功的盟可睐（Moncler）橱窗展示的关键要素时，Remo 透露：“我认为，在珍稀和独特的概念不断受到质疑的时刻，在盟可睐（Moncler），我们强调，卓越的最终标准是创造力。创造力能带来不同，激发人们的崇敬，唤起新的情景，并使一个身份在拥挤、高度连接的世界中脱颖而出。创造力来自研究、投资和无所畏惧。

图 3.11 香榭丽舍大街上的 Moncler 商店（经由 Moncler 许可）
Figure 3.11 Moncler Window on the Champs-Elysées (courtesy of Moncler)

It was rare to find a brand with deep roots in tradition. The goal was to bring the brand into the future starting from its origins. Our aim was to roll out a ‘global down jacket’ strategy all over the world. It was very important to communicate that Moncler is the jacket for all occasions. The brand’s evolution has become urban never renouncing the sporty spirit that is always present in Moncler’s soul. This is what we really wanted since the beginning of this fantastic ‘adventure’, starting with a great product and an outstanding history. Nothing to archive but definitely to develop.” (Derval, 2016)

Key 2 - Moncler Daring Displays

Moncler just opened a new store on the Champs-Élysées, the main shopping street in Paris, and gob smacked passersby and media with breathtaking visual merchandising and a treasure hunt, putting the outfit in scenes like no others. When talking about the key ingredients for a successful Moncler window display, Remo reveals “I think that in a moment when the concept of rarity and uniqueness are constantly questioned, at Moncler we strongly underline that the ultimate reference for excellence is creativity. Creativity makes the difference, inspires admiration, evokes new scenarios and makes an identity stand out in a crowded, hyper-connected world. Creativity comes from research, investments, and fearlessness.

这些主题是我们所有活动的基础，从视觉营销到活动，从沟通到展示和合作。因为如果说有一个因素决定了卓越，那就是在对时尚的实际关联性进行无休止的讨论之后，它在保持自身 DNA 基因不变的情况下，具有令人惊讶的能力。”

盟可睐（Moncler）通过将性能和优雅相结合，成为了羽绒服界的法拉利。重塑您的类别主要需要更好地了解客户的深层生理需求，明智的定位，以及一定的胆识，才能使您打破密码并重新定义游戏规则。

关键 3- 超越时尚

Remo Ruffini 也同意分享他对奢侈的定义：“我更喜欢优雅这个词，而不是奢侈。我经常引用萨拉·吉特里的一句法语：Le luxe est une affaire d’argent. L’élégance est une question d’éducation。”意思是“奢侈是金钱的问题，优雅是一个教育问题”。Remo 解释说，从他小时候起，父亲 詹弗兰科就一直是他的榜样：“他的优雅和风格真的影响了我。长大后我在美国住了一段时间，特别是在东海岸。我深深地被那些大家族低调的优雅和简朴所吸引。卡米洛特，海恩尼斯——美丽的肯尼迪家族和他们轻松、古典的气质，以及他们永恒的优雅和古典的接触的感觉。它是像阿涅利家族那样的意大利重要家族的美丽的美国版本。我认为，放松、轻松、别致的优雅深深影响了我。即使在今天，当我听到奢侈这个词时，对我来说，最直接的联想就是：表达出超越时尚的真实和个人优雅的东西。这也是我描述我的品牌盟可睐（Moncler）的方式。”这个品牌肯定能够发展并保持强大的 DNA。

These topics are at the base of all our activities, from Visual Merchandising to Events, and from Communication to Shows and Collaborations. Because if there is one element that determines excellence, out of the endless discussions on the actual relevance of fashion, it is its capability of surprising, while still maintaining its own DNA.”

Moncler became the Ferrari of the down jacket, by combining performance and elegance. Reinventing your category requires mostly a better understanding of clients’ deep physiological needs, a smart positioning, and a certain audacity that will enable you to break the codes and redefine the rules of the game.

Key 3 – Beyond Fashion

Remo Ruffini also agreed to share his definition of luxury: “I rather prefer the word elegance instead of luxury. In France there is a saying, from Sacha Guitry, that I often adopt: Le luxe est une affaire d’argent. L’élégance est une question d’éducation.” This means “Luxury is a matter of money. Elegance is a question of education”. Remo explains how since he was a child, his father Gianfranco has been a model to him: “his elegance and style really influenced me. Growing up I lived in the U.S. for a while, on the East Coast specifically. I was deeply attracted by the understated elegance and simplicity of the great families. Camelot, Hyannis – the beautiful Kennedys and their easy, classical flair, with their timeless elegance and classical touch. It was the beautiful American version of the important Italian families like the Agnelli’s. I think that relaxed, effortless, chic elegance influenced me a lot. Even today, when I hear the word luxury, for me the immediate association is: something that expresses authentic and individual elegance beyond fashion. This is also how I describe my brand Moncler.” The brand surely managed to develop and maintain a strong DNA.

3.6 接触奢侈品顾客：捷豹（Jaguar）案例 Reaching luxury shoppers: the Jaguar case

明确正确的奢侈品角色不仅有助于开发适合顾客群体的产品、服务和功能，如苹果（Apple）和宝马（BMW）案例所示，还有助于接触和连结顾客，如下文的捷豹（Jaguar）案例所示。

通过仔细分析品牌的角色，捷豹（Jaguar）成功超越 XK 车型的原定销售目标，同时节省了 30% 的全球宣传预算。

文森特，55 岁，大企业集团的董事会成员，年收入 15 万欧元以上。

他喜欢唐培里侬（Dom Perignon）的香槟，执迷于声望，个人主义至上。

产品的时尚程度是影响他做出购买决策的关键因素。

会议间隙，他会浏览金融时报等网站来跟踪手中股票的涨跌情况。

像其他决策者一样，文森特很难接近。

因此，捷豹（Jaguar）决定通过互联网与他互动，在他访问的金融网站上发布广告。

这种参与度是前所未有的，在金融网站上高达 45%，这意味着比在互联网上的平均表现高出 10 倍。销售紧跟其后（Derval,2018）。

更多了解他们的角色的日常活动和偏好，使得捷豹（Jaguar）能够节省 60% 的广告预算并和顾客联系紧密。

Identifying the right luxury persona will not only help develop suitable products, services, and features as we saw in the Apple and BMW case, it will also help reach customers and connect with them as we will see in this Jaguar case.
By carefully analyzing the brand Personas, Jaguar managed to exceed its sales objectives on the XK model while saving 30% on its global communication budget.
Vincent, 55 years old, is a member of the board of directors in a large corporate group and his revenues are over 150000 euros per year.
He prefers the champagne brand Dom Perignon and he is very much into prestige and individualism.
Fashion is a key factor in his purchasing decisions.
In between two meetings, he has a look at websites like the Financial Times to check the evolution of his stock options.
Like many deciders, Vincent is very difficult to reach.
So, Jaguar decided to interact with him over the Internet by posting advertising on the financial websites he is visiting.
The engagement rate was never seen before, with 45% on financial websites which means 10 times more than the average performance on the Internet. And the sales followed (Derval, 2018).
Knowing more about their personas' activities, agenda and preferences enabled Jaguar to save 60% on their global advertising budget and to connect deeply with their customers.

3.7 练习 Exercises

3A 土豪购物体验

根据土豪的角色（图 3.12）回答以下问题：

– 在购买服装和配饰时，这个角色在寻找什么？

– 他永远不会买什么？为什么？

– 他的采购动因是什么？

– 描述一个独特的购物体验，他最喜欢的品牌之一可以推荐给他。

用具体的例子说明你的答案。

3B 常春藤的角色

基于 Sephora（第 2 章）和 Kenzo（第 3 章）的案例，回答以下问题：

问题 1：运用角色框架描述辛迪的角色（图 3.13）。

问题 2 还有什么补充信息值得了解？为什么？

问题 3 你将如何获取这些缺失的信息？

您可以在DervalResearch代戈上获得角色框架：扫描二维码输入关键词 **设计3**。

小海龟幻觉的结果

A) 画面是移动的：你的眼睛在短波紫色和长波黄色之间来回移动，这似乎会使你的眼肌紧张，从而产生一种运动的错觉。

B）画面是不动的：你的眼睛毫不费力地在短波长的紫色和长波长的黄色之间来回移动，而不会绷紧你的眼肌，因此运动的错觉就不会发生。

3A The Tuhao shopping experience

Based on the persona of the tuhao (Figure 3.12) answer the following questions:

- What is this persona looking for, when shopping for clothing, and accessories?
- What will he never buy and why?
- What are his purchasing drivers?
- Describe a unique shopping experience one of his favorite brands could propose to him.

Illustrate your answers by specific examples.

3B Ivy's Persona

Based on the Sephora (in chapter 2) and Kenzo (in chapter3) cases, answer the following questions:

Question 1. Describe the persona of Ivy, using the persona framework (Figure 3.13).

Question 2. What additional information would be useful to know and why?

Question 3. How would you proceed to get this missing information?

You can get the persona framework on DervalResearch 代戈 wechat：scan the QRcode and type the keyword design3.

Results for the Baby Turtle illusion

A) The picture is moving: your eye is going back and forth between the short wavelength purple and the long wavelength yellow, which seems to tense your ocular muscle, so that an illusion of motion occurs.

B) The picture is not moving: your eye is going back and forth between the short wavelength purple and the long wavelength yellow effortlessly, without straining your ocular muscle, so that the illusion of motion does not occur.

图 3.12 土豪形象（经由 DervalResearch 代戈提供）
Figure 3.12 Tuhao Profile (Courtesy of DervalResearch 代戈)

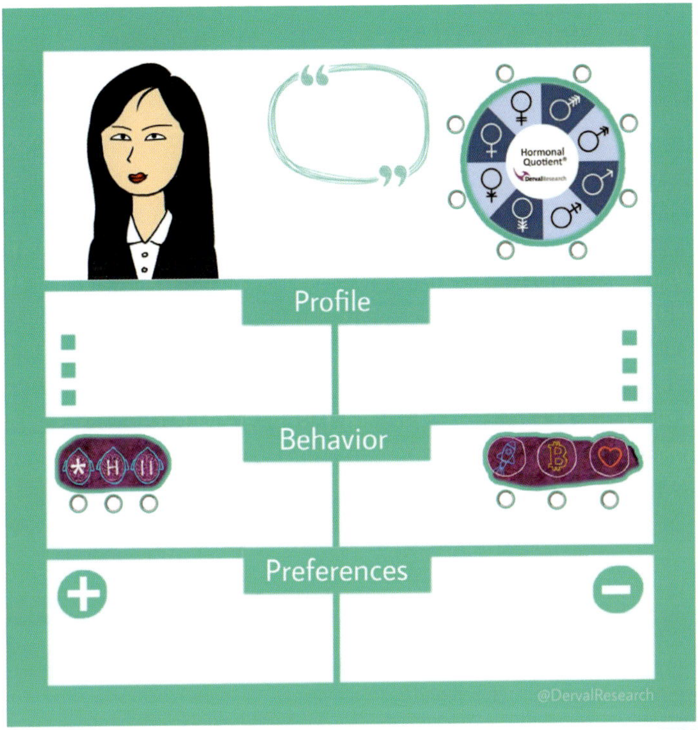

图 3.13 角色框架（经由 DervalResearch 代戈提高）
Figure 3.13 Persona framework (Courtesy of DervalResearch 代戈)

3

3.8 要点 Take-aways

- 通常根据年龄、性别或收入来细分群体的方法太模糊
- 角色有助于更有效地明确目标顾客群
- 了解角色有助于开发正确的设计、服务、功能，节省广告预算，并能将品牌与顾客联系起来。
- 要想独树一帜，就必须是本土化的和有针对性的
- 大胆的品牌设法深入了解他们的角色，从而使他们能够以高影响力承担适度的风险

- Usual marketing segments by age, gender, or revenue are too vague
- The Persona framework helps identify target customers in a more effective way
- Knowing the Persona will help develop the right design, services, features, save on the advertising budget, and connect the brand with its customers
- To be unique, an experience has to be local and targeted
- Daring brands manage to deeply understand their personas which allows them to take measured risks with a high impact

3.9 术语表 Glossary

角色

“角色（Persona）是一组具有相似行为模式的用户的代表。” (Mijač et al., 2018)

设计思维

“设计思维是设计师常用的一种过程，用于寻找复杂问题的解决方案，适应新的或不确定的环境，并创造新产品。” (Black et al.,2019)

重塑

“重塑一个品牌最关键的一步是回归到当初让它成为热门商品的东西。” (Agyemang et al.,2011)

Persona

“Persona is a representative of a group of users who share a similar behavioral pattern.” (Mijač et al., 2018)

Design Thinking

“Design thinking is a process commonly used by designers to find the solution to complex issues, navigate new or uncertain environments, and to create a new product.” (Black et al, 2019)

Revamping

“The most crucial step to revamping a brand is returning back to what made it a hot commodity from the beginning.” (Agyemang et al, 2011)

3

3.10 参考文献 References

Derval, D. (2022). The Right Sensory Mix: Decoding Customers' Behavior and Preferences. Springer Nature, ISBN: 978-3-662-63794-4.
Derval, D. (2018). Designing Luxury Brands: The Science of Pleasing Customers' Senses. Springer Business & Management, ISBN: 978-3-319-71555-1.
戴安娜 · 代尔瓦勒 (Diana Derval). 奢侈品品牌营销：创建 · 实施 · 案例 (Luxury Brand Marketing). 东华大学出版社 (2016), ISBN: 9787566910424.
Derval, D., Bremer, J. (2012). Hormones, Talents, and Career: Unlock your Hormonal Quotient. SpringerBriefs, ISBN: 978-3-642-25712-4.
Derval, D. (2011). Réussir son étude de marché en 5 jours. Eyrolles, ISBN: 978-2-212-54958-4.
DervalResearch (2020). The science of colors: From product to experience design. Iversity, Springer Nature Group. https://iversity.org/en/courses/the-science-of-colors
Murakami, I., Kitaoka, A., & Ashida, H. (2006). A positive correlation between fixation instability and the strength of illusory motion in a static display. Vision research, 46(15), 2421-2431.
Mijač, T., Jadrić, M., & Ćukušić, M. (2018, May). The potential and issues in data-driven development of web personas. In 2018 41st International Convention on Information and Communication Technology, Electronics and Microelectronics (MIPRO) (pp. 1237-1242). IEEE.
Black, S., Gardner, D. G., Pierce, J. L., & Steers, R. (2019). Design thinking. Organizational Behavior.
Agyemang, K. J. (2011). Athlete brand revitalisation after a transgression. Journal of Sponsorship, 4(2).

第四章 Chapter 4

激励购买 Motivating a Purchase

宝马香车 Luxury lifestyle

我们将在本章中看到，了解客户生活动机的设计师具有竞争优势。鲁布托（Louboutin）、积家（Jaeger-LeCoultre）和 TikTok 的案例将展示如何根据人物角色的动机制定差异化策略。国际单项体育联合会设计奖得主 Shane Lee 将分享如何进行一个有效的产品设计的秘诀。

目录

图 4.1 方方为 DervalResearch 代戈绘制的插画
Figure 4.1 Illustration by FangFang for DervalResearch 代戈

Designers who understand customers' motivation in life have a competitive advantage, as we will see in this chapter. The Louboutin, Jaeger LeCoultre and TikTok cases will show how to develop differentiated strategies based on the personas' motivation. iF Design awards winner Shane Lee will share tips on how to create an effective product design.

Content of this chapter

4.1 奢侈与地位追求 Luxury and Status Seeking

我们在 DervalResearch 代戈做的一项关于送礼的科学研究中，观察到了女性之间令人费解的差异。当我们要求她们指出她们一生中收到的最好的礼物时，一些人指着她们孩子的画（无疑是用更多的爱而不是天赋来完成的，但是，嘿，我们有什么资格去评判呢？此外，它是如此的可爱），而其他人用闪闪发光的眼睛展示给我们的，是他们最新的设计师包包。这并不一定意味着他们不太喜欢孩子的艺术作品，但就像这样，首要认知是，设计师包包是理想的礼物，尤其当它是最新款时（Derval,2018）。女性和男性一样，会更多地受到产前雄激素或雌激素的影响，这会对她们的行为和偏好产生影响（Derval & Bremer,2012）。在我们的实验中，提到孩子的画的女性更有可能是雌激素驱动的，而提到设计师包包的女性则更有可能是雄激素驱动的。

但是为什么女人一开始就那么喜欢包包呢？一些女性在争夺"大人物"时，她们选择的武器是时尚和奢侈以及碧池，又称"关系攻击"。

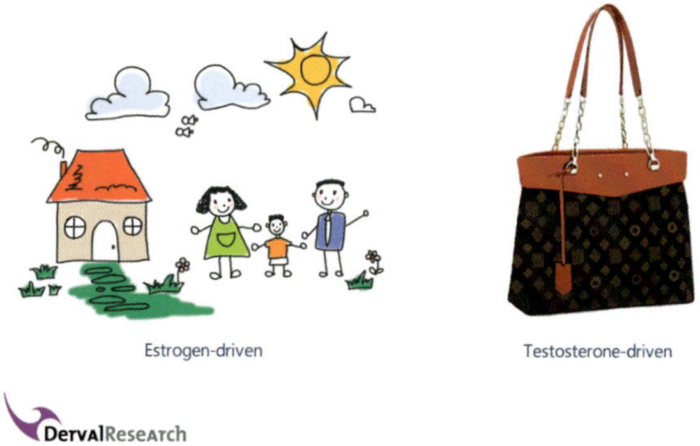

图 4.2 方方为 DervalResearch 代戈绘制的插画
Figure 4.2 Illustration by FangFang for DervalResearch 代戈

In an investigation we did at DervalResearch 代戈 on gifting, we observed bluffing disparities among women. When we asked them to point at the nicest gift they had ever received in their life, some pointed at their child's drawing (undoubtedly executed with more love than talent but, hey, who are we to judge? Plus, it is soooooo cuuuuuute) while others showed us, with glowing eyes, their latest designer bag. It doesn't necessarily mean that they like their children's artwork less but like that, top of the mind, a designer handbag is the ideal gift, especially when it is the latest one (Derval, 2018). Women like men can be more influenced by prenatal testosterone or estrogen and it has an impact on their behavior and preferences (Derval & Bremer, 2012). In our experiment, women who mentioned their child's drawing were more likely to be estrogen-driven and women who mentioned a designer bag were more likely to be testosterone-driven.
But why are women so much into bags in the first place? Some women are competing to get Mr Big and their weapon of choice is fashion and luxury and bitching aka "relational aggression".

关系攻击也可能被称为背后捅刀子或说坏话，它是指通过破坏或操纵与他人的关系来伤害某人——换句话说，她的声誉。雌性狒狒主要是为了在队伍中获得一个安全的地方而战斗，掠食者难以接近它们。有趣的是，研究表明，女性优势等级和相关特征，如较低的服从率，接触回避，或接触攻击，与产前雄激素效应（PAE）或产前睾酮（如果你更喜欢的话）的影响直接相关（Derval,2018）。

有意思的是，不是所有的有钱人都追求地位。看看挪威，这点尤为明显。这片孕育了维京人的土地蕴含丰富的石油，人均收入位列全球之首，但是挪威人的穿着却非常休闲。他们喜欢徒步旅行等户外活动，男人和女人都是身着不那么时尚的衣服：相似的普通牛仔裤和伐木工衬衫。最重要的是他们的衣服上没有任何表现地位的标志。另一方面，在其他国家，人们有时会把他们的全部工资去买一个金色或玫瑰金的 iPhone。对他们来说，地位是一种需要，奢侈品是一种必需品。最近的研究证实，有些人对某些多巴胺受体有更高的结合潜力，因此他们更看重社会地位——因为显示社会地位对他们来说是非常有益的。例如，法国人喜欢炫耀和寻求地位。社会和公司是等级森严的，一个人的权力往往通过汽车的大小或团队成员的数量来评估。因此，难怪许多奢侈品牌都来自法国——香奈儿（Chanel）、迪奥（Dior）、爱马仕（Hermès）、鲁布托（Louboutin）、索菲特（Sofitel）、马克西姆（Maxim's）、尚美（Chaumet）等，越来越多的奢侈品牌也正从中国崛起。

对于某些人而言，追逐地位其实是一种同吃饭喝水一样的生理需求。

Relational aggression might also be called back-stabbing or bad-mouthing and it is about harming someone by ruining or manipulating the relations to others – in other words her reputation. Female baboons would mostly fight in order to secure a safe place in the middle of the troop, less accessible to predators. Fascinatingly, it was demonstrated that the female dominance rank and associated traits such as lower submission rate, contact avoidance, or contact aggression were directly related to prenatal androgen effects (PAE) or the influence of prenatal testosterone if you prefer (Derval, 2018). Interestingly, not all rich people are looking for status. This becomes particularly obvious when checking countries like Norway. This land of Vikings has the highest income per inhabitant – due to the high concentration of petrol – still Norwegians dress up in a very casual way. While enjoying outdoor activities like trekking, men and women wear similar and not very fashionable outfits: jeans and "timberman shirts". Most importantly, they do not display signs of status.

On the other hand, in other countries people sometimes put their whole salary into a golden or rose gold iPhone. For them status is a need, and luxury products are a must-have. Recent research confirmed that some people have a higher binding potential on certain dopamine receptors and are therefore looking more for social status – as showing status is extremely rewarding to them. The French for instance like to show off and are seeking for status. The society and companies are very hierarchical and one's power is often assessed by the size of the car, or the number of team members. So, no wonder many luxury brands are coming from France – Chanel, Dior, Hermès, Louboutin, Sofitel, Maxim's, Chaumet – to name a few, and more and more are rising from China.

For some people status seeking is a physiological need, in the same way as eating and drinking.

4.2 购物者的动机：权力、性能和人气 Shoppers' Motivations: Power, Performance, and Popularity

如果追求地位与体内存在的 DRD2 / DRD3 多巴胺受体的数量有关，购物者会根据他们的深层动机表现出不同的行为和偏好：性能，权利或人气 (Derval, 2019)。

受性能驱使的购物者更关注产品的功能，而不是品牌本身。他们会青睐工程珠宝，例如积家（Jaeger-LeCoultre）手表或最新的戴森（Dyson）无叶风扇。这些客户根本不关心名人代言——对于每一款产品来说，“布丁好不好，吃了才知道”，他们的评论可以造就一个品牌，也可以毁掉一个品牌。

受权力驱使的购物者正在推动奢侈品行业，他们喜欢和大人物联系在一起。喜欢稀有和昂贵的产品，没有鞋跟够高，也没有材料够珍贵，因此偏爱鲁布托（Louboutin）或爱马仕（Hermès）。

受人气驱使的购物者不需要占主导地位，但他们喜欢属于一个群体。这些客户构成了奢侈品行业送礼顾客中不可忽视的一部分，他们将追随大众潮流——无论是围绕古驰（Gucci）配饰，迈克高仕（Michael Kors）手袋还是 TikTok 挑战。

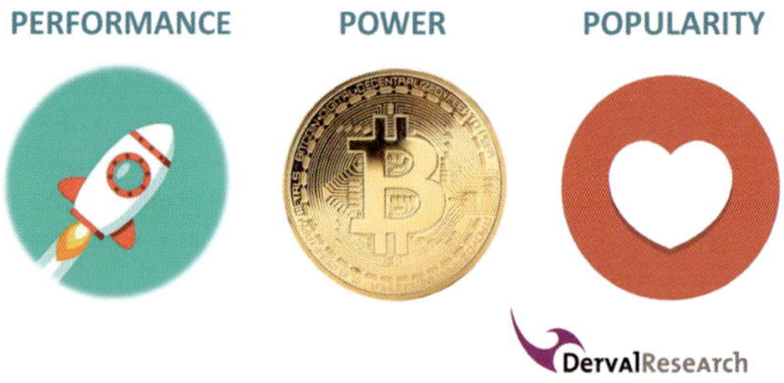

图 4.3 购物者的动机（经由 DervalResearch 代戈提供）
Figure 4.3 Shoppers' Motivations (courtesy of DervalResearch 代戈)

If status-seeking is linked to the number of DRD2/DRD3 dopamine receptors present in the body, shoppers display differentiated behavior and preferences based on their deep motivation: performance, power, or popularity (Derval, 2019).

Shoppers motivated by performance pay more attention to products' features than to brands themselves. Their favor will go to engineering jewels, such as Jaeger-LeCoultre watches, or the latest bladeless Dyson fan. These clients couldn't care less about celebrity endorsements – for each item the proof is in the pudding and their review can make or break a brand.

Shoppers motivated by power are driving the luxury industry and like to be associated with big wheels. Fond of rare and expensive products, no heel is high enough and no material is precious enough, hence a predilection for Louboutin or Hermès.

Shoppers motivated by popularity do not need to dominate but they like to belong to a group. These clients, who constitute a non-negligible part of the gifting customers in the luxury industry, will follow the pack – whether the trend is around Gucci accessories, Michael Kors handbags, or TikTok challenges.

4.3 鲁布托（Louboutin）：权力可以用英寸计量
Louboutin: Power can be measured in inches

借助鲁布托（Louboutin）案例，我们来看看这些购买驱动因素是如何成为识别奢侈品消费者的特征并超出他们的预期的关键。

让我们考虑一下高跟鞋。今天，高跟鞋是“行政女性”的象征，最初是为男性打造的。300 年前，波斯骑士为更好的骑马稳定性设计了这种鞋形。由于当时所有来自波斯的东西都很流行，路易十四和他的皇家宫廷都渴望穿上“It”鞋子。这很快变成了一场鞋跟测量比赛，因此路易十四不得不宣布他将是唯一一个允许穿红色高跟鞋的人来结束比赛。鲁布托（Louboutin）几乎就此诞生。高跟鞋是权力的代名词，从 1630 年开始被女性采用。

如今，运动鞋是权力的新象征。设计一双运动鞋不仅仅是把你的标志印在鞋子上。鲁布托（Louboutin）用极具魅力的“咬屁股”鲁比夏克（LoubiSharks）秀出了这一点。这种红色外底的运动鞋需要 67 个小时的制作时间，它会让收藏家和时尚引领者满意，从上海的女汉子到刚果男孩。漫画受到朋克风格的启发，创作了蓝色、灰色和紫色三种颜色。这是一次及时的发布，将解决敏感脚的角色，并为鲁布托（Louboutin）在亚洲和非洲开辟新的领域。

图 4.4 方方为 DervalResearch 代戈绘制的插画
Figure 4.4 Illustration by FangFang for DervalResearch 代戈

With Louboutin, let's see how these purchasing drivers are the key to identifying luxury shoppers' profiles and exceeding their expectations.

Let's consider high heels. Today a symbol of "executive women", high heels have originally been created for men. 300 years ago, Persian knights designed this shoe shape for better horse-riding stability. As all things coming from Persia were trending at that time, Louis XIV and his royal court were eager to wear the "It" shoes. This quickly turned into a heel-measuring contest, so that Louis XIV had to put it to an end by declaring he would be the only one allowed to wear red heels. The Louboutins were almost born. Synonymous with power, high heels were adopted by women starting 1630.

Today, sneakers are the new symbol of power. Designing a pair of sneakers is not just about putting your logo on shoes. Louboutin demonstrated it with the stunning and "bite-ass" LoubiSharks. The Sneakers with the red outersoles, that require 67-hours of work, will please collectors, and trendsetters from Shanghainese Nuhanzi to Congo Boys. Manga inspired with a punk touch, the creations come in blue, grey, and purple. A timely launch that will address the persona with sensitive feet and open new territories in Asia and Africa for Louboutin.

4.4 积家（Jaeger-LeCoultre）：性能和其他复杂功能 Jaeger-LeCoultre: Performance and other complications

客观上，由于时间是在智能手机上显示的，因此不需要手表。当然，如果有复杂功能的情况的话除外。除了是完美的商务礼物，手表对于某些人来说是性能的代名词。在航空公司和赛车驾驶员的推荐下，手表品牌竞相赢得终极标杆：太空征服。从欧米茄（Omega）的月球漫步限量款使月球漫步名垂青史，到泰格豪雅（Tag Heuer）携手特斯拉（Tesla）和太空探索技术公司（SpaceX）征服火星，手表和它们的复杂性令人着迷。复杂性是奢侈手表的附加值。它们的范围从天文台表到超复杂腕表（Grand Complication），如测速仪，一种能改善重力感的陀飞轮，再到月球周期管理——以便始终找到理发的吉祥时间(Derval, 2018)。

拥有400多项专利的积家（Jaeger-LeCoultre）设计了最小和最薄的机械表，自1833年以来一直被誉为真正的钟表匠——提供江诗丹顿（Vacheron），万国表（IWC），百达翡丽（Patek Philippe）或爱彼（Audemars Piguet）等品牌背后的技术。这也是历峰集团（Richemont）在其奢侈品牌投资组合中增加了该瑞士品牌（性能的代名词）的原因之一。

受性能驱使的购物者对产品功能的反应比对品牌的反应更大。重要的是要能够展示技术优势，例如通过为其他主要参与者制造产品（如积家案例），或通过在极端条件下的管理能力，在太空征服、一级方程式或电子竞技模式下。运动装品牌安德玛（Under Armour）刚刚与维珍银河宇航员合作制作服装而得分。

解码客户就是识别他们的动机。通过他们的准则和策略，品牌会吸引更多的特定类型的客户。

Objectively, since time is displayed on smartphones, there is no need for a wristwatch. Except of course, if there are complications. In addition to being the perfect business present, a watch is for some synonymous with performance. Recommended by airline and racing pilots, watch brands compete to win the ultimate benchmark: space conquest. Between Omega that immortalized the moonwalk with its limited Moonwalker edition and Tag Heuer associating with Tesla and Space X to conquer Mars, watches and their complications fascinate. Complications are the added-value of luxury watches. They range from a chronometer to grand complications like a tachymeter, a tourbillon that improves the sense of gravity, to lunar cycle management – in order to always spot the auspicious time for a haircut (Derval, 2018).

With over 400 patents behind their belt, Jaeger- LeCoultre, who designed both the smallest and the thinnest mechanical watch, has been admired since 1833 for being a true watchmaker – providing the technology behind brands like Vacheron, IWC, Patek Philippe, or Audemars Piguet. One of the reasons why Richemont added the Swiss maison, synonymous with performance, to its luxury brands portfolio.

Shoppers motivated by performance respond to product features more than to brands. It is important to be able to demonstrate technological superiority, for instance by manufacturing for other main players like in the Jaeger-LeCoultre case, or by managing prowess in extreme conditions, in space conquest, Formula 1, or e-sports mode. Sportswear brand Under Armour just scored points by co-creating Virgin Galactic spacecrew outfits.

Decoding clients is a matter of identifying their motivation. Through their codes and strategies, brands will attract more of certain types of clients.

4.5 与 Shane Lee 一起设计一个出色的设计 Devising an outstanding design with Shane Lee

人们对权力、性能或人气度产生了兴趣，并对每种类型的顾客都有特定的要求。Shane Lee，工商管理硕士，三星 (Samsung) 和 TCL 等领先品牌的设计主管，将就如何提供有效而不过分夸张的设计给我们提供建议。

第一步，对你的设计充满信心

“这将增加成本，这将是非常困难的，因为需要定制的组件……等。” Shane 承认，但是他坚持说，“我总是收到来自研发的投诉，但我相信在我 20 年的设计经验的基础上我可以作出正确的决定。”

第二步，使你的设计突出但不怪异

Shane 解释道：“几乎每个人都把相机模块放在角落里——这是主流的设计布局。因为这样的布局使用的资金及研发成本较少。” 但 Shane 警告说：“如果你遵循风险更低的主流设计，人们将不会记得你的设计。”

第三步，在设计和成本之间取得平衡

“如果一切都听从研发和采购团队的话，就不会有好的设计。一个好的设计师需要了解成本结构，并权衡自己的设计。” Shane 提到：“我把每一分钱都当成自己的钱运用在设计中。”

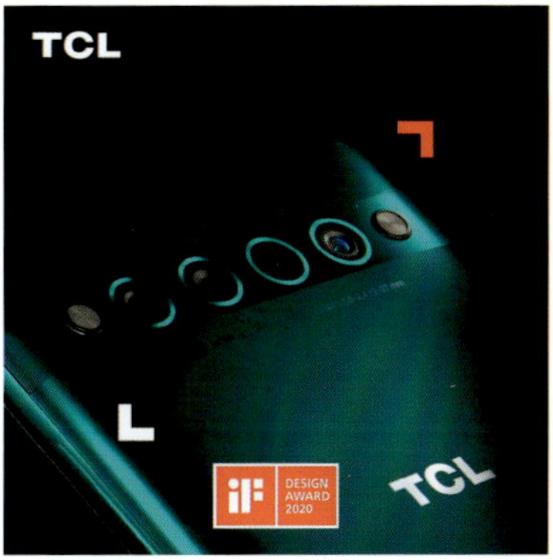

图 4.5 TCL iF DESIGN AWARD（经由 TCL 和 iF DESIGN 代戈许可）
Figure 4.5 TCL iF DESIGN AWARD (courtesy of TCL and iF DESIGN)

People are into power, performance, or popularity, and each type of customer has specific requirements. Shane Lee, MBA, head of design for leading brands like Samsung and TCL, will give us advise on how to deliver effective design without overkilling it.

Step 1 Be confident in your design

“It will add cost, it will be very difficult due to customized component required...etc.” admits Shane but he insists: “I always got complaints from R&D but I believe I can make the right decision in design based on my 20 years design experience.”

Step 2 Make your design outstanding but not odd

“I don’t like to give a feeling of copying the iPhone design or the Samsung design,” explains Shane, “everybody puts the camera module in the corner – it is the mainstream design layout”, because it requires less investment and less R&D effort but Shane warns:“People will not remember your design if you follow the mainstream design, with the lower risk.”

Step 3 Find a balance between design and cost

“There is no good design if everybody listens to the R&D and the sourcing teams. A good designer needs to understand the cost structure and make the trade-off for their own design” highlights Shane and he adds: “I treat every penny as my own money.”

第四步，规划一种进化的设计语言

Shane 补充了共享设计语言的重要性：“来自同一家族的一系列产品可以使用 CMF(颜色、材料和饰面) 的级别和成本来调整从高阶到低阶的设计语言。”

第五步，在生产过程中注意细节

“如果你不注意，每一个细节都会影响到整个设计”，Shane 建议在设计生产和上市之前都不要放松警惕。

Shane Lee 的设计在拉斯维加斯的消费电子展和 iF 设计奖上广受好评。

iF 设计奖始于德国汉诺威，拥有强大的工业技术，70 年来一直在选择世界各地的最佳设计，包括产品、包装、建筑和用户体验等领域。我有幸在成都新开的 iF 设计中心上了一节大师课。Joe 欢迎我们来到这个海外的第一个 iF 设计中心并分享一些见解。正如上海是时尚之都一样，成都正在成为中国的设计之都。每年，在成都创意设计周期间，来自世界各地的优秀设计师都会来到四川的首府。成都 iF 设计中心已经邀请了名创优品、高仪、小米等知名品牌进行首轮展览，期待更多来自不同领域、不同品牌的优秀设计。作为创意设计的中心，成都 iF 设计中心以其设计咖啡厅、设计商店、联合办公空间、会议室和活动空间吸引了来自世界各地的顶级设计师、设计工作室和参观者。除了令人垂涎的 iF 设计奖，iF 现在每年都为学生设立的 iF 设计人才奖，以及 iF 社会影响奖，以促进社会的可持续发展。您可以在 iF 的官方网站 iF WORLD DESIGN GUIDE 上了解细节。

Step 4 Plan an evolutive design language

Shane adds on the importance of a shared design language: “A series of products from the same family can use the level of CMF (Color, Material, and Finish) and cost to adapt the design language from high-tier to low-tier.”

Step 5 Keep an eye on details during production

As “every detail can impact the whole design if you don't pay attention to it”, Shane recommends not to put the guards down before the design is produced and hits the market.

Shane Lee’s designs have been acclaimed from CES in Las Vegas to iF DESIGN AWARD.

iF DESIGN AWARD started in Hanover Germany with a strong industrial know-how and have been consistently selecting the best designs worldwide, in various fields like product, packaging, architecture, and UX, for 70 years. I was lucky to give a masterclass at the newly opened iF design center in Chengdu. Joe, who welcomed us in this first iF DESIGN center overseas, accepted to share some insights. In the same way Shanghai is a capital of fashion, Chengdu is becoming the design city of China. Every year, during the Chengdu Creative & Design Week the best designers from around the world are coming to Sichuan’s capital. iF DESIGN center Chengdu has invited well-known brands such as MINISO, Grohe, and Xiaomi for its first round of exhibitions, and many more outstanding designs from a variety of fields and brands are to be expected. As a hub for creative design, iF DESIGN center Chengdu attracts top designers, design studios, and visitors into design, from all over the world with its design café, design shop, co-working spaces, meeting rooms, and event spaces. In addition to the coveted iF DESIGN AWARD, iF proposes now every year the iF DESIGN TALENT AWARD, set up for students, as well as the iF SOCIAL IMPACT PRIZE, established to promote a sustainable development of the society. To apply, you can find all details on iF official website: iF WORLD DESIGN GUIDE.

4.6 TikTok 成功背后的秘密：人气和归属感 The Secret Behind TikTok Success: Popularity and Belonging

TikTok 的巨大成功不仅仅是因为小城市缺乏娱乐。社交平台正是充分利用了人类的竞争和约会本能。

那些社交能力强，因此在认知移情上得分高的人，会无意识地模仿他人，采取同样的姿势，像揉鼻子一样的姿势，甚至是同样的面部表情，让自己看起来更熟悉、更迷人（Chartrand&Bargh,1999）。

这种“变色龙效应”可能与“镜像神经元”的发现有关，镜像神经元是一种让我们模仿同类的生理机制。人们希望，模仿最成功的人也能让我们成功 (Derval, 2018)。因此，TikTok 成功了 (Derval,2022)。

字节跳动（Bytedance）社交平台也用于约会节目。例如，女性会评估一个男性，考虑他跳舞时的波动不对称性——这意味着所有由于健康问题、遗传学、寄生虫或协调而导致的问题正在被扫描。与唱歌和面部结构一样，身体运动是睾酮循环水平的良好预测指标（Derval & Bremer,2012），也是产前睾酮水平的良好预测指标（Manning & Taylor,2001）。上半身（颈部、腹部）的动作似乎决定了“好”和“坏”舞者之间的区别 (Derval,2018)。

这一切都解释了字节跳动（Bytedance）不仅在中国本土取得了巨大成功，而且在 100 多个国家取得了巨大成功，从印度（其最大的市场）到泰国。

The considerable success of TikTok is not just due to the lack of entertainment in smaller cities. The social platform just makes the most of human competing and dating instincts.

People who are very social, and therefore score high on cognitive empathy, imitate others unconsciously, adopting the same posture, the same gesture like rubbing the nose, and even the same face expression, making themselves look more familiar and engaging (Chartrand & Bargh, 1999).

This “chameleon effect” can be linked to findings on the “mirror neurons”, a physiological mechanism that makes us imitate our peers. The hope is that imitating the most successful individuals might lead us to become successful as well (Derval, 2018). Hence the success of TikTok. (Derval, 2022).

Bytedance social platform is also used for dating reperage purposes. Women for instance will evaluate a male, consider his fluctuating asymmetries when he is dancing – that means all the little deviations in movements' symmetry imputable to health issues, genetics, parasites, or coordination problems are being scanned. Body movement, in the same way as singing and face structure, are good predictors of circulating levels of testosterone (Derval & Bremer 2012), and of prenatal testosterone (Manning & Taylor, 2001). The movements of the upper body (neck, abdomen) seem to make the difference between “good” and “bad” dancers (Derval, 2018).

This all explains the tremendous success of Bytedance not only in its home country China but in over 100 countries from India – its biggest market – to Thailand.

中国品牌在各个行业都处于领先地位：在 C2C 销售平台上，淘宝对 Ebay；在电信基础设施方面，华为对思科；在社交媒体应用方面，微信对 Facebook；在在线支付方面，支付宝对 Paypal。如果说这些品牌大多忙于在中国各省扩张，那么其中一些品牌就会效仿海尔、锦江和李宁的成功经验，开始出口自己的产品和服务。这是拥有 TikTok 的字节跳动（Bytedance）的案例。即使在估值方面，目前全球估值最高的初创企业字节跳动（Bytedance）在滴滴之前，在任何西方公司之前 (Andrews,2019)。在中国经济格局中，抖音的定位是由外国投资者拥有的，是独创性的，但与华为（由其员工拥有的公司）一样，目标是在中国国内外采取积极的扩张战略 (Fannin,2019)。字节跳动（Bytedance）正在考虑成为音乐流媒体服务的更大竞争者，并通过推出自己的智能手机来加强其社交媒体地位。把硬件和人工智能结合起来，如果结果好的话，在其他人失败的地方，全力以赴，听起来就像是一个智能世界统治计划 (Derval,2022)。字节跳动(Bytedance) 有办法做到这一点，因为他们一贯提供必需的功能。我们可以引用 TikTok 自动面部 / 身体关键点检测，以方便短视频编辑，基于标签和观看行为的“为您”人工智能推荐，录制视频评论的“反应”功能，以及从收购的 music.ly 音乐平台继承的著名“二重唱”功能，非常适合竞赛和匹配 (KrASIA,2018)。受欢迎的人喜欢归属，喜欢成为一个群体或潮流的一部分。对于以他们为目标的品牌来说，总是推出价格实惠的配饰，并与 TikTok 上的热门人物合作是至关重要的。

Chinese brands lead the way in various industries: TaoBao against Ebay in C2C selling platforms, Huawei against Cisco in telecom infrastructure, Wechat against Facebook in social media applications, or Alipay against Paypal in online payment. If most of these brands are busy expanding throughout the Chinese provinces, some of them follow the success story of Haier, Jinjiang, and Li-Ning and start exporting their products and services. This is the case of Bytedance with TikTok (aka Douyin). Even in terms of valuation, the highest valued start-up in the world is now Bytedance right before Didi, and before any Western company (Andrews, 2019). Owned by foreign investors, Douyin and TikTok positioning in the Chinese economical landscape is original but like Huawei – a company owned by its employees – the objective is to have an aggressive expansion strategy in and outside of China (Fannin, 2019). Bytedance is considering becoming a bigger contender in the music streaming service and strengthening its social media position by launching its own smartphone. Combining hardware and AI, and if it turns out well, where others failed, landing in all hands sounds like a smart world domination plan (Derval, 2022). Bytedance has the means to achieve this as they consistently deliver must-have features. We can cite TikTok's automatic facial/body keypoint detection facilitating short video editing, the "for you" AI-powered recommendations based on tags and watching behavior, the "react" feature to record comments on a video, and the famous "duet" feature to sing along another video, inherited from acquired Musical.ly music platform, and ideal for competing and match-making purposes (KrASIA, 2018). People motivated by popularity love to belong, to be part of a group, or a trend. It is critical for brands who target them to always propose affordable accessories and to collaborate with hot TikTok influencers.

4.7 练习 Exercises

4A 神秘之城阿姆斯特丹游戏（45 分钟）

在探索阿姆斯特丹，解开一个谜团的同时，你将以自己的速度发现神经科学、用户体验（UX）和奢侈品管理的最新进展。

无需事先了解就可以跟随这一教育冒险！首先，扫描二维码并输入关键词**设计 4（或神秘 1）**，然后按照每集结尾的说明进行操作。

图 4.6 City of Mystery Amsterdam（刘治淳和袁于然为 DervalResearch 代戈绘制的插图）
Figure 4.6 City of Mystery Amsterdam (Illustration by Zhichun Liu and Yuran Yuan for DervalResearch 代戈)

4A City of Mystery Amsterdam Game (45min)

While exploring Amsterdam, and solving a mystery, you will discover at your own pace the latest advances in neuroscience, user experience (UX), and luxury management. No prior knowledge is required to follow this educational adventure!
To start, scan the QRcode and enter the keyword design4 (or mystery1) and follow the instructions at the end of each episode.

然后回答以下测验：

问题 1- 克拉是以什么名字命名的

a、法国城市巴卡拉，位于南希和斯特拉斯堡之间，自 1764 年以来以其水晶眼镜而闻名

b、希腊语“keration”（角豆之果），因为每颗角豆的重量正好为 200 毫克，被用来称钻石的重量

c、巴基斯坦的卡洛特村，位于旁遮普，一个著名的钻石抛光服务地区

问题 2- 为了导航，维京人曾使用

a、地球的磁场，因为光在它上面的折射会在蓝天上划出一条金色的线，叫做“海丁格尔刷形象”，他们可以跟随

b、磁罗盘，中国人在公元前 206 年发明的磁罗盘，最早用于按照风水原理来安排物品

c、他们的船！叫做德拉卡（“龙”）

问题 3- 味蕾有助于感知甜味、酸味、咸味、鲜味和苦味。如果我们计算一下 1 平方厘米舌头上味蕾的数量，有些人会发现有 10 个味蕾，有些人则会有

a、11 个味蕾　b、100 个味蕾　c、1100 个味蕾

问题 4- 人造钻石在 ____ 中构成

a、化学气相沉积　b、高压高温　c、以上两者都可以

Then answer following Quiz:
Question 1 - Carat is named after
a. the French city of Baccarat, located between Nancy and Strasbourg, and famous for its crystal glasses since 1764
b. the Greek “keration” (“fruit of the carob”), as each carob bean weighs exactly 200 milligrams and was used to weigh diamonds
c. the Karot village in Pakistan, located in the Punjab, an area well known for its diamond polishing services
Question 2 - To navigate, Vikings were using
a. the magnetic field of earth, as the light's refraction on it would draw a golden line in the blue sky called “the Haidinger’s Brush”, they could follow
b. the magnetic compass invented by Chinese people in 206BCE and first used to harmoniously arrange items following feng shui principles
c. their ship! called drakkar (“dragon”)
Question 3 - Taste buds help perceive sweet, sour, salty, umami (satiety), and bitter. If we would count the number of taste buds on 1 cm^2 of tongue, in some people we would find 10 taste buds and, in some people
a. 11 taste buds
b. 100 taste buds
c. 1100 taste buds
Question 4 - Man-made diamonds are grown
a. by CVD, Chemical Vapor Deposition
b. by HPHT, High Pressure High Temperature
c. either way

问题 5- 近视眼的人（他们从近处看比从远处看得清）看 ____ 更放松

a、蓝色，因为这种短波颜色正好照射到视网膜前部的焦点所在位置

b、红色,因为这种长波的颜色正好照射到视网膜的后部,它们的焦点就在那里

c、BTS 歌曲

问题 6- 被光吸引的个体被称为

a、古北区的

b、虫子

c、偏振光

问题 7- 成瘾是由位于身体和大脑中的阿片受体的数量、结合电位和类型来定义的。卡帕阿片受体对咖啡和 ___ 有反应

a、糖，β－内啡肽，脂肪和蛋白质

b、花生、薄荷、尼古丁和绿茶

c、它们的名字

问题 8- 一种有限的本体感觉，也就是定位自己身体的能力，会让你不擅长

a、网球中的抛接和发球

b、骑自行车和系鞋带

c、以上都是

Question 5 – Near-sighted people (they see better from close by than from a distance) are more relaxed by

a. Blue, as this short-wavelength color hits the front of the retina exactly where their focal point is located

b. Red, as this long-wavelength color hits the back of the retina exactly where their focal point is located

c. BTS songs

Question 6 - Individuals attracted to light are called

a. Palearctic

b. bugs

c. polarotactic

Question 7 - Addiction is defined by the number, binding potential and type of opioid receptors located in the body and brain. Kappa opioid receptors respond to coffee and to

a. sugar, beta-endorphin, fat, and protein

b. peanuts, mint, nicotine, and green tea

c. their name

Question 8 - A limited sense of proprioception, which is the ability to locate your own body, will make you bad at

a. juggling and serving in tennis

b. riding a bike and tying shoelaces

c. all of the above

问题 9- 不会被一种类似于汽油的气味激怒的人

a、可以驾驭任何香味

b、喜欢水果味

c、喜欢淡淡的花香

问题 10- 伦勃朗是著名画家的 ____

a、昵称

b、名字

c、姓

完成了吗？查看解决方案见第 77 页。

4B 如何销售夹克？

你是一个建议买一件昂贵的冬季夹克的小贩。你会用什么样的方式来说服喜欢权力的顾客？喜欢性能的顾客？喜欢人气的顾客？

完成了吗？查看解决方案见第 77 页。

4C 游艇案例

谢赫·阿里想为他的游艇购买雷松（Raymarine）或菲利尔（Flir）导航系统。他的游艇宽敞舒适，还配有船长。那么他的选购要求和标准会是什么呢？他追求什么呢？替代产品又会是什么呢？

完成了吗？查看解决方案见第 77 页。

Question 9 - People not irritated by a scent like gasoline

a. can handle any fragrance

b. prefer fruity scents

c. prefer subtle floral scents

Question 10 - Is Rembrandt the famous painter's

a. nickname

b. firstname

c. family name

Done? Check the solution page 77.

4B How to sell a jacket?

You are a vendor proposing an expensive winter jacket. With what type of argument, will you convince a target customer into power? a target customer into performance? a target customer into popularity?

Done? Check the solution page 77.

4C The Yacht case

Sheikh Ali wants to buy a Raymarine/Flir navigation system for his yacht. The yacht is a big yacht piloted by a captain. What are Sheikh Ali requirements and buying criteria for this material? What is he looking for? What are the substitution products?

Done? Check the solution page 77.

练习题 4A 的测验答案

问题 1 b、问题 2 a、问题 3 c、问题 4 c、问题 5 a、问题 6 c、问题 7 b、问题 8 c、问题 9 a、问题 10 b

练习题 4B 的解决方案

喜欢有权有势的角色会喜欢最昂贵、最新的夹克衫，用难以获得的面料制作而成，并采用复杂的工艺。

喜欢性能的角色会很乐意听到夹克在极端天气条件下的防护等级，以及切口、口袋或层如何提供最佳功能。

喜欢人气的角色在电影明星的上一部电影中穿着一模一样的冬季夹克时，他可能会留下深刻的印象。或者，它是下一季紧跟潮流的安全选择。

练习题 4C 的解决方案

谢赫·阿里（Sheikh Ali）想要给同行留下深刻印象，他正在寻找一件独特的物品来炫耀。任何能在游艇上展示的独特而又酷的小玩意都是潜在的竞争者，如一个巨大的电视屏幕，一个金色的酒吧，等等。

图 4.7 方方为 DervalResearch 代戈绘制的插画
Figure 4.7 Illustration by FangFang for DervalResearch 代戈

Quiz Answers for exercise 4A:

Question 1: b, Question 2: a, Question 3: c, Question 4: c, Question 5: a, Question 6: c, Question 7: b, Question 8: c, Question 9: a, Question 10: b

Solution to exercise 4B

A persona into power will like the most expensive, latest jacket, made with difficult to obtain fabrics and using a complex and intricate process.

A persona into performance will appreciate to hear about the jacket's protection level in extreme weather conditions and how the cut, pockets, or layers can provide the optimal function.

A persona into popularity might be impressed to learn that a popular actor was wearing the exact same winter jacket in his last movie. Or that it is the safe option for the coming season to stay on trend.

Solution for Exercise 4C

Sheikh Ali wants to impress his peers and is looking for a unique item to show off. Any unique and cool gadget that can be showcased in the yacht is a potential competitor. Think of a giant TV screen, a golden bar, etc.

4.8 要点 Take-Aways

- 对一些人来说，追求地位是一种生理需要，就像吃饭或喝酒一样
- 对于有地位的人来说，时尚奢侈品是“必备品”
- 棘手的部分是创造出有影响力的设计，而不至于过度破坏
- 在不同的地理区域，人们渴望爬上社会阶梯和炫耀自己的言论会有所不同
- 需要不同的策略和消息来转换角色，这取决于角色是权力、性能还是人气

- For some people, status seeking is a physiological need, in the same way as eating or drinking
- For people into status, fashion and luxury items are a “must-have”
- The tricky part is to create impactful designs without overkilling it
- The eagerness to climb the social ladder and to show-off statement pieces will vary between geographical areas
- Differentiated strategies and messages are required to convert personas depending whether they are into power, performance, or popularity

4.9 术语表 Glossary

变色龙效应

“变色龙效应是指一个人无意识地模仿对方的姿势、行为举止、面部表情和其他行为，从而使自己的行为在当前的社会环境中被动和无意地改变以与他人相匹配。” (Chartrand et al., 1999)

地位

奢侈品可以提供地位，但“信息性礼物也强烈地受到地位和地位追求的驱动，地位感更有可能维持虚拟社区”。(Lampel, 2007)

气质

“气质被定义为生物根源，主要是天生的，是人格和其他个性特征的基础。” (Nęcka, 2013)

Chameleon effect

“The chameleon effect refers to nonconscious mimicry of the postures, mannerisms, facial expressions, and other behaviors of one's interaction partners, such that one's behavior passively and unintentionally changes to match that of others in one's current social environment.” (Chartrand et al., 1999)

Status

Luxury can provide status but “informational gift giving is also strongly driven by status and status seeking, and that status sentiments are more likely to sustain virtual communities.” (Lampel, 2007)

Temperament

“Temperament is conceptualized as the biologically rooted, mostly inborn, foundations for personality and other individual traits.” (Nęcka, 2013)

4.10 参考文献 References

Derval, D. (2022). The Right Sensory Mix: Decoding Customers' Behavior and Preferences. Springer Nature, ISBN: 978-3-662-63794-4.

Derval, D. (2019, November 19). Luxe : Vos clients sont-ils motivés par le pouvoir, la popularité ou la performance ? Harvard Business Review. https://www.hbrfrance.fr/chroniques-experts/2019/11/28509-luxe-vos-clients-sont-ils-motives-par-le-pouvoir-la-popularite-ou-la-performance/

Derval, D. (2018). Designing Luxury Brands: The Science of Pleasing Customers' Senses. Springer Business & Management, ISBN: 978-3-319-71555-1.

戴安娜 · 代尔瓦勒 (Diana Derval). 奢侈品品牌营销：创建 · 实施 · 案例 (Luxury Brand Marketing). 东华大学出版社 (2016), ISBN: 9787566910424.

Derval, D., Bremer, J. (2012). Hormones, Talents, and Career: Unlock your Hormonal Quotient. SpringerBriefs, ISBN: 978-3-642-25712-4.

Derval, D. (2011). Réussir son étude de marché en 5 jours. Eyrolles, ISBN: 978-2-212-54958-4.

iF Design. (2021). TCL 10 PRO. iF WORLD DESIGN GUIDE. https://ifworlddesignguide.com/entry/285490-tcl-10-pro

Chartrand, T. L., & Bargh, J. A. (1999). The chameleon effect: the perception–behavior link and social interaction. Journal of personality and social psychology, 76(6), 893.

Manning, J. T., & Taylor, R. P. (2001). Second to fourth digit ratio and male ability in sport: implications for sexual selection in humans. Evolution and human behavior, 22(1), 61-69.

Andrews, M. (2019, December 4). TikTok shock. CKGSB Knowledge. https://english.ckgsb.edu.cn/knowledges/bytedance-tiktok-shock/

Fannin R. (2019, September 13). The Strategy Behind TikTok's Global Rise. Harvard Business Review. Retrieved from https://hbr.org/2019/09/the-strategy-behind-tiktoks-global-rise

Porter, J. (2019, July 29). TikTok owner ByteDance confirms plans to produce a smartphone. The Verge. Retrieved from https://www.theverge.com/circuitbreaker/2019/7/29/8934846/tiktok-smartphone-bytedance-smartisan

KrASIA Writers. (2018, June 28). A closer look at TikTok's expansion overseas. KrASIA. Retrieved from https://kr-asia.com/a-closer-look-at-tik-toks-expansion-overseas

Lampel, J., & Bhalla, A. (2007). The role of status seeking in online communities: Giving the gift of experience. Journal of computer-mediated communication, 12(2), 434-455.

Nęcka, E., & Hlawacz, T. (2013). Who has an artistic temperament? Relationships between creativity and temperament among artists and bank officers. Creativity Research Journal, 25(2), 182-188.

第五章 Chapter 5

捕捉趋势 Capturing Trends

各色各样 In all shapes and sizes

在本章中，我们将解读如何捕捉趋势。因此，荷尔蒙商数将揭示一个珍贵的预测工具。特斯拉的创新路线图将显示如何预测未来趋势。杨设计将分享一些关于时尚潮流的内幕消息。兰蔻（Lancôme）、香奈儿（Chanel）和李宁（Li-Ning）的案例将突出大胆在设计中的重要性，以启动大趋势。

目录

图 5.1 Aubery 为 DervalResearch 代戈绘制的插画
Figure 5.1 Illustration by Aubery for DervalResearch 代戈

In this chapter we decode how trends can be captured. For that matter, the Hormonal Quotient® will reveal a precious predictive tool. Tesla innovation roadmap will illustrate how it is possible to anticipate future trends. And Yang Design will share some insider tips on fashion trends. Lancôme, Chanel, and Li-Ning cases will highlight the importance of audacity in design, in order to start major trends.

Content of this Chapter

5.1 预测工具：荷尔蒙商（HQ）
Predictive Tools: the Hormonal Quotient®

预测哪种趋势会成功的最佳方法是充分理解每一类角色，他们的产品偏好、生理情况和动机。例如，在购买太阳镜时，追求原创产品的角色可能被产品新颖的形状、材料和色彩所吸引，而寻求地位的角色可能会喜欢金色的框架。关键就是理解用户身上每一种时尚主张背后的动机：为什么他们会喜欢某件产品？一旦我们明确了问题的答案，就能非常容易地预测同类型的用户会购买哪一款新产品。

人们的购买决策也会受到诸多生理因素的影响。个人的体型、脸型都将决定他会选择什么形状的眼镜，个人的视力情况会决定他喜欢什么颜色。此外，个人肌肤对材质的敏感程度和承受疼痛的能力会影响他选择的产品材质和重量——太重的镜架有些人戴着会感觉很疼，所以即使他们喜欢彰显身份的金色镜架，也可能会因为太重而放弃。

图 5.2 荷尔蒙商环形图（经由 DervalResearch 代戈提供）
Figure 5.2 Hormonal Quotient® wheel (courtesy of DervalResearch 代戈)

The best way to predict which trends will be successful is to perfectly understand each persona, their product preferences, physiology, and motivations. For instance, when buying sunglasses, a persona looking for originality might be attracted to novel shapes, materials, and colors, whereas a persona looking for status might enjoy golden frames. The key is to understand the motivation behind each fashion statement: why does this person wear this item? Once we have a clear answer it makes it very easy to predict which items the same person will pick in the new collection.

Many physiological factors will influence the purchasing decision. The body shape and face shape will influence the shape of the selected glasses, the vision will determine the favorite colors, the sensitivity to texture and pain will influence the selected material and its weight – some people for instance cannot stand heavy frames as it is too painful, so even if they like to display status, the golden frames might be too heavy for them.

DervalResearch 代戈在超过 25 个国家进行研究，根据上万名参与者的生理数据，开发了强大且能细分人群生理状况的预测工具：荷尔蒙商（HQ）。我们发现人们凭借味蕾、内耳的毛发细胞、眼部的视锥视柱细胞和皮肤感受器对产品和服务产生知觉——这都同荷尔蒙商相关。了解了消费者的荷尔蒙商，我们就可能预测他们喜爱的颜色、口味、气味、形状、材质和声音。

基于对 50 多个不同目标群体的研究，包括高级行政人员、家庭主妇、创业者、采购经理和意见领袖，我们开发了荷尔蒙商并归纳出 8 种不同的状况。只要恰当地使用消费者的荷尔蒙商数据，公司就能根据他们不同的生理状况，为市场设计并向消费者传达正确的体验，预测他们的感官知觉、购物行为、产品喜好。这样公司能节约不少成本，无需再开展传统的调查，既周而复始又不便宜。这一切只要明确消费者的概况和荷尔蒙商即可。

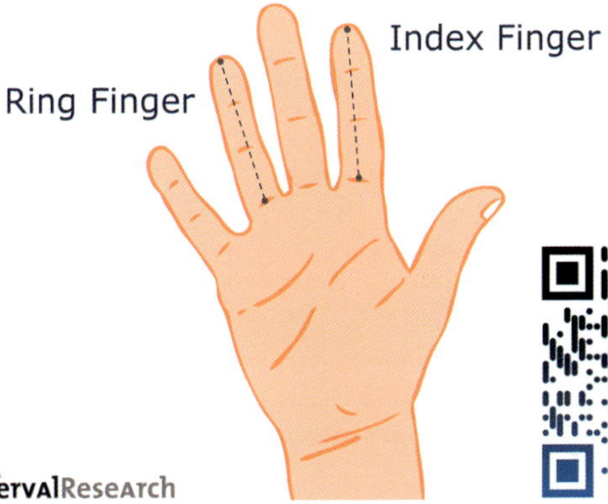

DervalResearch

图 5.3 荷尔蒙商测试（经由 DervalResearch 代戈提供）
Figure 5.3 Hormonal Quotient® Test (courtesy of DervalResearch 代戈)

Drawing upon thousands of measurements in over 25 countries, DervalResearch 代戈 developed a powerful and predictive biological segmentation tool: the Hormonal Quotient® (HQ). We have discovered that people's perception of products and services – via their taste buds, hair cells in the inner ear, rod and cone cells in the eyes, and skin sensors – is linked to their Hormonal Quotient®. Knowing consumers' HQ makes it possible to predict not only their favorite colors, but also their preferred tastes, smells, shapes, textures, and sounds.

The Hormonal Quotient® tool was developed by studying over 50 target groups, including top executives, housewives, entrepreneurs, purchasing managers, and opinion leaders. From this we were able to determine eight different Hormonal Quotient® profiles. Applied properly, these profiles will allow firms to design and deliver the right consumer experience across their markets in a very cost-effective way, predicting consumers' sensory perceptions, purchasing behavior, and product preferences based on their biological profiles. Companies no longer need to conduct traditional, recurrent, and costly surveys. They just have to identify the profile and Hormonal Quotient® (HQ) of their consumers once.

我们被某些特定的名人吸引也是由于生理和体内化学反应作用。男人和女人都会受到雄性激素和雌性激素的影响，打娘胎里就是这样。受到每种激素程度决定了我们的思想、面貌以及行为 (Derval & Bremer, 2012)。

DervalResearch 代戈通过测量超过 25 个国家的成千上万人的表现，确认了 4 种类型的男人和 4 种类型的女人：

你可能受雌性激素影响更多，比如维多利亚·贝克汉姆和布拉德·皮特；受雄性激素影响更多，比如艾玛·沃森和阿诺德·施瓦辛格；特别受到雄性激素影响，如安吉丽娜·朱莉和斯蒂夫·乔布斯；或者两者平衡（意味着受到雌性激素和雄性激素的影响相当），如麦当娜和乔治·克鲁尼，以及我们第三章提到的宝马 BMW 的用户角色模型加里。

我们身体的每一个部位都可以作为生物标志物，就像一块化石，证明产前激素水平一样。

现在你明白了为什么有着安吉丽娜·朱莉般方下巴的雄激素驱动的女性要动手术来获得维多利亚·贝克汉姆般的三角下巴。这并不是说要改善外表——我认为美有很多种类型——但更多的是看起来像另一种多态性，可能在某个特定的国家更受欢迎，因为这种情况很少见。如果你像我一样过于敏感，不适合做手术，你可以去旅行，找到一个更重视你身体特征的地方（Derval，2018）。

你可以通过 www.dervalresearch.com 或扫描二维码来检测你的荷尔蒙商 (图 5.3)。

The way we are attracted to certain celebrities also is physiological and chemical. Both men and women are influenced by both testosterone and estrogen, while in the womb. The level of exposure to each hormone shapes the way we think, we look, and behave (Derval & Bremer, 2012).

Based on measurements performed on thousands of people in over 25 countries, DervalResearch 代戈 , identified 4 types of men and 4 types of women:

You can be more influenced by estrogen, like Victoria Beckham and Brad Pitt, more influenced by testosterone like Emma Watson and Arnold Schwarzenegger, very influenced by testosterone like Angelina Jolie and Steve Jobs, or balanced (which means equally influenced by estrogen and testosterone) like Madonna and George Clooney, and Gary our BMW persona in chapter 3.

Each and every part of our body acts as a biomarker, like a fossil testifying of the levels of prenatal hormones.

Now you understand why many testosterone-driven women with Angelina Jolie-like square jaws are having surgery to get a Victoria Beckham-like triangular chin. It is not so much about improving appearance – I think there are many types of beauty – but more about looking like another polymorphism maybe more popular in a given country because it’s rare. If, like me, you are too sensitive for surgery, you can always travel to find a place where your physical traits are better valued (Derval, 2018).

You can check your Hormonal Quotient® at www.dervalresearch.com or by scanning the QRcode (Figure 5.3).

5.2 特斯拉的创新路线图 Innovation Roadmap with Tesla

观察消费者——他们穿什么、有什么行为都是清晰的线索，让我们能预测他们以后的行为，或推测他们对于新的环境或产品会有什么反应。让我们看看创新品牌特斯拉（Tesla）是如何完美地运用这一点的。那么，为什么预测是可能的呢？它们之所以有可能，是因为同一背景下的同一原因会导致同样的结果。当然，困难在于很好地理解原因、背景和影响。怎么做？主要是通过观察。例如，在古代，希腊人和罗马人做鸟占卜。鸟兆是指观察一只鸟，从它的行为，它飞行的方式，或它说话或不说话的方式，你知道有什么不对劲。因此，首先，人们倾向于嘲笑这种做法，但最近，已经证明，鸟类，至少某些类型的鸟类，有能力看到地球磁场。这意味着他们可能能看到和听到我们大多数人看不到的东西。你可以试着观察自己。假设你认识一个对游戏感兴趣的人。为了观察这个角色的购物行为，你可以观察每月、每周或季节性的行为。有时人们会在光棍节或特殊场合买更多的东西，或者你可以检查每月的行为。如果我知道一个人物喜欢游戏，我不知道他这个月会买什么，但我确信，他会买一些东西。所以你可以在不同的层次上做预测：如果你知道这个人喜欢玩游戏，你肯定知道，他们肯定会买东西，你的预测就会准确。如果你更了解玩家，并且根据玩家的口味对游戏市场（包括下一版本）有了深入的了解，你就可以准确地预测他们将投资哪些配件、哪些资源或哪些游戏。

Observing consumers – what they wear, how they behave – gives us clear clues on how they will behave in the future or when confronted with a new situation or new products. Let's see how innovative brand Tesla perfectly understood how to do that. So how come predictions are possible? They are possible because the same cause in the same context will lead to the same effects. The difficulty of course is to nicely understand the cause, the context and the effects. How to do that? Mostly through observation. So already in ancient times, the Greek and the Romans were for instance doing bird augury. Bird augury is about looking at a bird and from its behavior, the way it is flying, or the way it is talking or not talking, you know that something is wrong. So first, people tended to mock this way of doing but recently, it has been demonstrated that birds, certain types of birds at least, have the ability to see the magnetic field of the earth. Which means that they can probably see and hear things that most of us cannot. You can try to observe yourself. Let's say you know someone who is into gaming. To observe the shopping behavior of this persona you can look at the monthly, weekly, seasonal behavior. Sometimes people buy more for Single's Day or for special occasions or you can check the monthly behavior. If I know someone is into gaming, I don't know what they will buy this month but I know for sure, that they will buy something. So, the prediction you can do it at different levels : if you just know that the person is into gaming, you know for sure, they will buy something for sure and your prediction will be accurate. If you know the gamer even better and you have a deep understanding of the gaming market – including the next releases – based on the gamer's taste, you can predict exactly which accessories, which resources or which games they will invest in.

当然，要做出准确的预测，还需要具备一定的人物角色和背景知识 (Derval, 2022)。

奢侈品结合创新会引起强烈的反响。跟随着香奈儿（Chanel）的步伐，汽车品牌特斯拉（Tesla）是当前受争议的奢侈品品牌。闻名于世的贝宝（Paypal）创始人埃隆·马斯克大胆设计了奢华的电动汽车。就连流行的《汽车族》杂志也不得不承认特斯拉（Tesla）是本年度最值得称道的汽车。这个来自加利福尼亚的电子运动型汽车让尼桑（Nissan）、丰田（Toyota）和雪佛兰（Chevrolet）等其他力图说服司机改开电动汽车的品牌黯然失色，这些品牌推出的电动汽车像一个个会发出尖锐噪音的移动铁盒子。马斯克是第一个相信优雅和现代化能结出双赢果子的人。他本人曾涉足太空旅行、创办太阳能公司，以及参考最近的人工智能道德规范的制订——大概是为了确保以后的汽车不会吃了司机？——现在他成为了与维珍集团创始人理查德·布兰森，阿里巴巴集团主席马云同一级别的新意见领袖。人们痴迷于他的故事，封他为“现实版钢铁侠”，他甚至还真的去客串了钢铁侠电影 (Derval, 2016)。

It requires of course a certain expertise of the persona and of the context to make an accurate prediction (Derval, 2022).
Luxury and innovation create a tumultuous combination. Following the steps of Chanel, car brand Tesla is the current controversial luxury brand. Elon Musk, already known for creating Paypal, dared to design a luxury electric car. Even the popular magazine Motortrend acclaimed Tesla as the car of the year. The California-based electric sports car brand clearly outshined the Nissan, Toyota, and Chevrolet trying to convert drivers to electric cars with squeaking boxes on wheels. Musk was the first to believe in a winning combination of elegance and modernity. With his involvement in space travel, solar energy, and more recently artificial intelligence ethics – maybe to make sure cars do not start biting their drivers? – he became a new opinion leader, joining the ranks of Richard Branson of Virgin, or Jack Ma of Alibaba. People are so fascinated by his story that he gained the surname of “real life ironman” and he was even featured in the movie (Derval, 2016).

图 5.4 方方为 DervalResearch 代戈绘制的插画
Figure 5.4 Illustration by FangFang for DervalResearch 代戈

一个创新产品路线图强大的框架，通常由市场和研发部门共享，将有助于设计您的项目。在创新路线图的上部，我们确定了新技术、趋势、新竞争对手和替代产品，以及未来几年可能影响目标客户的风险。根据行业的不同，分析这些外部因素的时间线可能在 5 到 20 年之间变化。就房地产而言，20 年听起来不错，因为该公司还忙于设计和建造房屋。在创新路线图的下部，我们考虑到外部因素，组织内部创新项目和产品发布。根据外部风险和机遇调整我们的项目将有助于提高创新命中率。让我们将这个工具应用到高速列车案例中，重点关注特斯拉（Tesla）(图 5.5)。我们可以认为 2030 年是正确的时间线。

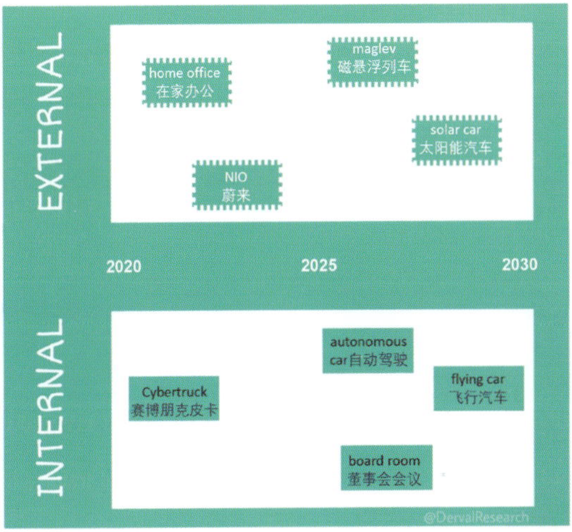

图 5.5 特斯拉创新路线图 (经由 DervalResearch 代戈提供)
Figure 5.5 Tesla Innovation Roadmap (courtesy of DervalResearch 代戈)

An innovation roadmap – powerful framework often shared by marketing and R&D departments – will help you design your projects. In the upper part of the innovation roadmap, we identify new technologies, trends, new competitors and substitution products, as well as hazards likely to influence the target customers in the coming years. The timeline for analyzing these external factors may vary between 5 and 20 years depending on the industry. In the case of real estate, 20 years sounds good as the firm is also busy with designing and building the houses. In the lower part of the innovation roadmap, we organize the internal innovation projects and product launches taking into account the external factors. Aligning our projects to external risks and opportunities will help increase the innovation hit rate. Let us apply this tool to the high-speed trains case, with a focus on Tesla (Fig. 5.5). We can consider that the horizon of 2030 is the right timeline.

对于特斯拉（Tesla）来说，发生的外部事件可能是：

• 家庭办公：有些人已经在家工作了。如果他们中的许多人到 2025 年都在家工作呢？

• 磁悬浮：包括磁悬浮列车和无轨电车在内的高速列车发展迅速。

• 蔚来：以蔚来为首的新一代电动汽车，拥有更长的自主性（得益于一个巧妙的电池替换系统），可能会危及特斯拉，甚至在中国市场之外。

• 太阳能汽车：奇怪的是，埃隆·马斯克 (Elon Musk) 还没有进入太阳能汽车市场，因为他的家人在很多屋顶上安装了太阳能电池板，而特斯拉 (Tesla) 提出了 Powerwall，一种巨型家用电池，以避免电力短缺。像 Aptera 这样的公司正在启动新的太阳能电动汽车市场，应该会有更多的公司跟进。

特斯拉的创新可能包括：

• 赛博卡车 (Cybertruck): 特斯拉 (Tesla) 新款 SUV 的防弹窗户和奇形怪状的外形让人惊讶。这个巨大的屋顶是为了支撑一个可选的太阳能电池板而设计的，这个电池板很难嵌入到其他较小的特斯拉车型中 (Lambert, 2020)。

• 自动驾驶汽车：对这两家加州品牌来说，第一次没有驾驶的测试并没有产生决定性的结果，但我们可以预计，到 2025 年，它们将推出改良版。

• 董事会会议室：蔚来汽车从一开始就意识到，自动驾驶会改变车内的活动，并重新安排座位区域。我们可以期待新的特斯拉车型建立在豪华轿车的基础上，提供相当于一个董事会会议室的空间，供出差的商务人士使用。

For Tesla external events occurring could be:

• Home-office: Some people are already working from home. What if many of them work from home by 2025?

• Maglev: High-speed trains, including magnetic levitation trains and trams without tracks, are developing quickly.

• NIO: A new generation of electric cars led by NIO, with a longer autonomy (thanks to an ingenious battery replacement system), could endanger Tesla and this even outside of the Chinese market.

• Solar cars: Weird that Elon Musk did not enter the solar car market yet as his family equips many roofs with solar panels and Tesla proposes Powerwall, a giant home battery to avoid electricity shortages. Players like Aptera are kicking-off the new solar electric vehicle market and more should follow.

Tesla's innovations could include:

• Cybertruck: Tesla new SUV surprised by features like bullet-proof windows and its singular shape. The large roof was designed to support an optional solar panel that was to tricky to embed in other smaller Tesla models (Lambert, 2020).

• Autonomous cars: First tests without a driver were not conclusive for the Californian brands but we can expect a refined version by 2025.

• Board room: NIO got it from the start, that autonomous driving would change activities conducted in a car and re-arranged the seating areas. We can expect new Tesla models built on a limousine model and hosting the equivalent of a board room, for business people on the move.

• 飞行汽车：等待瞬间运输——马斯克发起的超级高铁项目现在由维珍公司开发——飞行汽车可能是下一个很酷的事情。随着电动飞机的发射，天空中已经布满了无人机，我们可以预计到 2030 年，会有一些会飞的汽车，尤其是 SpaceX 公司的团队设计的 (Derval, 2022)。

为了创建一个坚实的创新路线图，你需要了解角色和趋势。例如，是什么阻止了一些人拥有电动汽车？对于那些饱受细颗粒物折磨的哮喘患者来说，这听起来很不可思议，但有些人确实喜欢汽油的味道：

- 1832 年，100% 的汽车（= 第一辆汽车！）是电动的
- 1900 年，28% 的汽车是电动的
- 2018 年，5% 的汽车是电动的
- 自 1920 以来，福特以汽油为燃料的汽车接管了汽车工业

这是一个很好的例子，说明了为什么要改变一个市场，真正的购买驱动因素是关键。建议速战速决，以促进汽油爱好者对电动汽车的采用：在他们的电动车上挂一棵汽油味的树 (Derval, 2022)。

关于特斯拉，你现在明白了，例如，一个被汽油味激怒的人更可能选择一辆电动车，但同样也可能选择一辆蔚来。

• Flying cars: Waiting for teleportation – the Hyperloop project initiated by Musk is now being developed by Virgin – a flying car is probably the next cool thing. As electric planes are being launched, and the sky is already full of drones, we can anticipate some flying cars by 2030, especially by the teams working on SpaceX (Derval, 2022).

To create a solid innovation roadmap, you need to understand personas and trends. For example, what prevents some people from embracing electric vehicles? As incredible as it sounds for asthmatic folks who suffer from fine particles, some people do love the smell of gasoline:

- in 1832, 100% of the cars (= the first car!) were electric
- in 1900, 28% of the cars were electric
- in 2018, 5% of the cars were electric
- since 1920, Ford killed it with gasoline-fueled vehicles

This is a great example showing why understanding the real purchasing drivers is key in order to change a market. Suggested quick-win: Hanging gasoline scented trees in electric cars (Derval, 2022).

Concerning Tesla, you understand now that for instance a persona who is irritated by the smell of gasoline is more likely to opt for an electric car – but might equally select a NIO.

5

5.3 兰蔻，成功之声 Lancôme, the sound of success

许多知名品牌都有拗口的名字。我们可以想到施华洛世奇（Swarovski），谷歌（Google）或保乐力加（Pernod-Ricard）。这并没有阻止它们取得成功。那么怎样才能取一个好的品牌名称呢?

从戴森（Dyson）到香奈儿（Chanel），在许多情况下，使用姓氏是首选的解决方案，除非它听起来不太好，比如兰蔻 Lancôme 的案例。

当 Armand Petitjean 决定退出与香水制造商 François Coty（他后来创立了这个同名品牌）的商业合作伙伴关系，创建自己的美容产品品牌时，他认为自己的姓氏，以及他刚刚想到的家乡圣卢普（Saint-Loup），听起来都不是很女性化。因此，创始团队选择了附近的森林和城堡，名为兰科姆（Lancosme）。受到像旺多姆（Vendôme）这样的奢侈品大牌的启发：品牌兰蔻（Lancôme）诞生了。

Petitjean 想要一个听起来像法语的名字，以区别于美国竞争对手，该名称因带有字母“l”，“m”和“ô”而听起来圆滑，并因“k”发音而听起来有力。对徽标和包装的感官研究表明，声音和形状的一致组合可以改善顾客体验。关键是使声音与品牌密码保持一致。例如，施华洛世奇（Swarovski）听起来像水晶一样尖锐而冰冷。值得强调的是，在《财富》500 强公司保护的品牌名称中，“k”是一个代表过多的字母。“K 趋势”始于美国 80 年代的肯德基（KFC），可口可乐（Coca-Cola），并一直延续到星巴克（Starbucks）或家乐福（Carrefour）。

Many famous brands have tongue-twisting names. We can think of Swarovski, Google, or Pernod-Ricard. And it didn’t prevent them from becoming successful. So what makes a good brand name?
Using the family name is in many cases the best solution - from Dyson to Chanel – unless it doesn’t sound so good, like in the Lancôme case.
When Armand Petitjean decided to quit the business partnership with perfume maker François Coty (who since then founded the eponym brand) to create his own beauty products brand, he considered that neither his surname, nor his hometown of Saint-Loup, sounded feminine enough. The founding team therefore opted for a nearby forest and castle called Lancosme. Inspired by luxury big names like Vendôme: the brand Lancôme was born.
Petitjean wanted a French-sounding name to differentiate from American competitors. The name sounds round with the letters “l”, “m”, and “ô”, and powerful with the sound “k”. Sensory research on logos and packaging shows that a congruent combination of sound and shape improves the customer experience. The key is to align the sound with the brand codes. For instance, Swarovski sounds sharp and cold like a crystal. It is worth highlighting that “k” is an over-represented letter in the brand names protected by Fortune 500 companies. The K- trend started in the US in the 80’s with KFC, Coca-Cola, and continues with Starbucks or Carrefour.

即使对说法语的各位来说，保乐力加（Pernod-Ricard）的发音也并不容易，尤其是在喝了几杯之后。这个品牌变成了 Ricard。同样，路易威登（Louis Vuitton）在其首字母缩写 LV 下更为人所知。在截断 -ing 和首字母缩写 -ing 之后，昵称 -ing 成为趋势 (Derval, 2018b)。

例如，在中国，兰蔻领先的美容精华液 Génifique 已更名为“小黑瓶”。值得一提的是，兰蔻（Lancôme）对比小黑瓶（Génifique）的声音强度是后者的两倍，并且令人钦佩地加强了黑瓶已经笨重的尺寸。正在崛起的化妆品品牌 SK-II 看到了其明星产品 SK-II 面部护理精华，绰号为“神仙水”，母公司宝洁（Procter & Gamble）对此有很好的反应，注册并保护了这个美好的昵称。

当听起来好听的时候，最好是为品牌使用姓氏，例如鲁布托（Louboutin）或米其林（Michelin），或者名字，比如蒂芙尼（Tiffany）、马克西姆（Maxim's）或梅赛德斯（Mercedes）。不要吝啬使用神话里的神，像爱马仕（Hermès）信使，或者女神（DS）。宝马（BMW）等地名，字面意思是巴伐利亚汽车公司（Bayerische Motoren Werke）也很好用。

如果您的角色或目标顾客，以及品牌密码是明白的，找到品牌名称是简单的例行公事。好消息是，只要产品是好的，用户总会找到方法去命名它，最重要的是购买它。

Pernod-Ricard is not easy to pronounce, even for French-speaking folks, particularly after a few glasses. The brand became Ricard. In the same way, Louis Vuitton is better known under its initials LV. After truncation and acronymization, nicknaming is trending (Derval, 2018b).

In China for instance, Génifique – Lancôme's leading beauty serum – has been renamed “the little black bottle”. It is worth mentioning that Lancôme and Génifique sound twice as strong and strengthen admirably the already massive dimension of the black flask. The raising cosmetics brand SK-II saw its star product SK-II facial treatment essence nicknamed “magic water” and the mother company Procter & Gamble had the good reflex to trademark and protect this nice nickname.

When it sounds good, best is to use a surname for the brand, like Louboutin or Michelin, or a first name, think of Tiffany, Maxim's, or Mercedes. Why go without a mythology god like Hermès, the messenger, or a goddess/ DS. Place names like BMW, literally the Bavarian motors (“Bayerische Motoren Werke”) also work well.

If your personas or target customers, and brand codes are clear, finding the brand name is a simple formality. Good news is that as long as the product is good, users will always find a way to name and most importantly to buy it.

5.4 香奈儿 5 号：从趋势到图标 Chanel N° 5: From Trend to Icons

香奈儿（Chanel）依靠品牌同玛丽莲·梦露等偶像的联系建立了享誉国际的名声。当这个女演员透露说睡觉时只需一滴香奈儿 Chanel N° 5 香水就像穿了睡袍一样时，顿时在媒体上掀起轩然大波，香水销量也随之暴涨。香奈儿（Chanel）由此登上国际时尚舞台，并从此再没离开过（Derval，2018a）。

可可·香奈儿本人性格叛逆，崇尚自由精神。她为女装带来了裤装设计，以及迄今为止专用于男装内裤的针织棉布，此等创举令女装焕然一新。香奈儿还帮助女性重新认识了自己。她创造的现代设计从某种意义上来说是对传统服装的反抗。当 Chanel 开始成为中产阶级专属品牌——满足知名人士妻子的需求时，卡尔·拉格菲尔德成功地让香奈儿（Chanel）重新回归了这一优势。秉承着法式戏剧的情节发展规则，这故事的确在有戏剧冲突的地方令人愈发激动。香奈儿（Chanel）敢于创新，敢于打破规则。

香奈儿的精神依然存在。如今，许多国际明星，比如集格莱美制作人、嘻哈歌手、创作人和时装设计师等多重身份于一身的法瑞尔·威廉姆斯等国际巨星都穿着巴黎品牌的服装，并搭配一些非常经典的作品，如将长珍珠项链融入他们的街头风格。

图 5.6 方方为 DervalResearch 代戈绘制的插画
Figure 5.6 Illustration by FangFang for DervalResearch 代戈

Chanel actually built its international success by associating its brand with icons like Marilyn Monroe. When the actress revealed that she was sleeping with just a drop of Chanel N°5 as her nightgown, it made a huge sensation in the media and the sales of the perfume exploded, putting Chanel on the international fashion scene, and the brand never left since then (Derval, 2018a).
Coco Chanel, rebel and free spirit, innovated in women's wear, introducing pants and cotton jersey, fabric used only for men's underpants so far. Chanel helped women reinvent themselves, and Coco with the help of her modern designs was somehow fighting tradition. Karl Lagerfeld managed to give Chanel an edge back, when the brand started to appeal to "bourgeoises" only – understand big wheels' wives. Chanel dared to innovate and break the rules.
And Chanel spirit is still alive. Nowadays international stars like multi Grammy winning producer, rapper, singer, songwriter, and fashion designer Pharrell Williams are seen wearing the Paris brand and incorporating some very classic pieces like the long pearl necklace into their street style.

5.5 与杨设计识别趋势 Spotting trends with YANG DESIGN

时尚和设计发展迅速，在网络上有大量信息可用，因此识别相关趋势非常复杂。幸运的是，YANG DESIGN 的获奖设计师黄晓靖曾参与波音公司的设计，并每年发布一本流行趋势书，重点介绍如何有效地识别趋势。

他们为 00 后设计的个人护理包(图 5.7)最近获得了 Pentawards 金奖。

第一步：探索未来生活方式

为了探索中国未来的生活方式，我们从经济、政治、设计、文化、社会和技术等 6 个方面进行了调查。

第二步：设计专家访谈

与来自 6 个设计领域的专家进行对话：媒体、金融、艺术家、家具设计师、产品设计师、材料设计师。

图 5.7 YANG DESIGN 设计的个人护理包装(经由 YANG DESIGN 提供)

Figure 5.7 Personal care package designed by YANG DESIGN (courtesy of YANG DESIGN)

Fashion and design are moving fast, and a lot of information is available online, so that identifying relevant trends is complex. Luckily, award-winning designer Xiaojing Huang from YANG DESIGN, who worked on Boeing designs and releases every year the popular "China Design Trends Report" highlights how to effectively spot trends.

Their personal care package (Figure 5.7), designed for the Generation Z, recently won the Pentawards Gold Award.

Step 1: Explore future lifestyle

To explore the future lifestyle in China, the 6 following fields are investigated: economy, politics, design, culture, society, and technology.

Step 2: Interview design experts

Follows conversations with experts from 6 design fields: media, finance, artist, furniture designer, product designer, material designer.

第三步：确定宏观趋势

“符号学研究（通过对符号的观察和分析）有助于产生 4 个宏观趋势”，黄晓靖解释道，她也是年度“中国设计趋势报告－中国设计趋势展区”展览的策展人，在上海设计展期间，共接待了 7 万名参观者。

第四步：优化预测

然后对预测进行细化（风格、色彩、图案、材质和表面处理）。

用户体验研究往往研究用户为什么这样做，需要应用心理学、社会学、CMF 设计等跨学科知识。

第五步：可视化人物角色和场景

最后一步是将“有用户体验场景和人物角色的目标消费群体”视觉化，黄晓靖以她为领先品牌设计的创新空调为例：“用户希望从视觉上感受到强烈的打击，所以空调的前部应该设计多个出风口，并增加吹风面积和吹风角度，产品整体造型也应体现出强劲的风力、线条的动感。”

图 5.8 YANG DESIGN CMF（色彩、材料与表面处理）创新实验室（经由 YANG DESIGN 提供）
Figure 5.8 YANG DESIGN CMF (Color, Materials and Finish) Innovation Lab (courtesy of YANG DESIGN)

Step 3: Identify macro trends

“A semiotic research (via the observation and analysis of signs) helps then generate four macro trends” explains Xiaojing who is also the curator of the annual “China Design Trends Report - China Design Trends Gallery” exhibition, which welcomed 70,000 visitors during Design Shanghai.

Step 4: Refine the forecast

The forecast is then refined (style, color, pattern, material and finish).

User experience research often studies why users do so, and needs to apply interdisciplinary knowledge such as psychology, sociology, CMF design and so on.

Step 5: Visualize personas and use cases

The last step is about visualizing the “ target consumer group with UX scenarios and personas” and Xiaojing takes the case of the innovative air-conditioning she designed for a leading brand: “ Users want to feel the strong blow visually, so the front of the air-conditioning should be designed with multiple air outlets, and increase the blowing area and angle; the overall shape of the product should also reflect the strong blowing force with dynamic lines”.

5.6 李宁：一切皆有可能 Anything is possible by Li-Ning

106 枚是李宁职业生涯中获得的令人印象深刻的金牌数。他可能厌倦了穿着不合身的外国品牌，并在 1990 年决定推出自己的品牌，使中国运动员能够自豪地穿上一个民族品牌。

在耐克和阿迪达斯的较量中，只有一个赢家：李宁！

体操王子李宁先生 30 年前创立的这个品牌令人震惊地卷土重来，向大家展示了它是如何做到的。中国品牌如何从 30 亿元的亏损中恢复过来？确切地说，它形成了一种趋势。让我们分析一下发生了什么，从中吸取教训。这位体操王子在中国南部靠近越南的山区长大，之后与同样来自 56 个少数民族之一的广西壮族自治区的奥运会奖牌得主结婚（Li-Ning, 2021）。

作为在 2008 年北京夏季奥运会上点燃火炬的榜样，李宁不怕挑战耐克（Nike）和阿迪达斯（Adidas）等全球品牌。他的设想是，中国人会购买它们，因为它们比廉价的服装更舒适。高端的本地仿冒者将更多地关注材料的昂贵使用，同时也忽略了技术和可穿戴性。全球品牌失去了一点光彩，所有品牌都倾向于稳妥行事。

106 is the impressive number of golden medals Li Ning collected throughout his career. He was probably tired of wearing ill-fitting foreign brands, and decided in 1990 to launch his own griffe to enable Chinese athletes to proudly wear a national brand.

In the battle between Nike and Adidas, there can only be one winner: Li-Ning!

The brand created 30 years ago by the Prince of Gymnastics Mr. Li Ning made an astounding come back, showing everybody how it is done. How did the Chinese brand recover from a 3 billion yuan loss and literally set a trend? Let's analyze what happened and learn from the best. Married to a fellow Olympic medal champion also coming from the Guangxi Zhuang Autonomous Region and Zhuang – one of the 56 Chinese ethnic minorities – the prince of gymnastics grew up in this mountain area in the South of China, near Vietnam (Li-Ning, 2021).

A role model, the man who lit the torch at the 2008 summer Olympic games in Beijing, was not afraid to defy global brands like Nike or Adidas. His vision was that people of China would buy them because they were more comfortable than cheap lookalikes. Premium local copycats would focus more on expensive use of materials, missing also the mark on technology and wearability.

图 5.9 Aubery 为 DervalResearch 代戈绘制的插图

Figure 5.9 Illustration by Aubery for DervalResearch 代戈

李宁在研发方面投入了相当大的精力，并在“韦德之道”系列中成功地将EVA 中底和橡胶外底结合起来，中间有一层碳纤维层，以吸收冲击，鞋跟缓冲可以反弹。

李宁共同打造了“韦德之道”系列，并根据 NBA 球员的需要定制了这双鞋，以便他能在比赛中自豪地穿着。首字母“ZZDG”代表了他的儿子 Zaire 和 Zion，他的侄子 Dada，女友 Gabrielle 的名字以及数字 92 和 12116 号分别代表他的祖母、孩子和侄子的年龄。韦德很欣赏李宁，他们都对这次特别合作有共同的看法 (Dwyer, 2021)。

最近的“中国李宁”系列在天猫上售罄，包括蔡依林和贝拉哈迪德在内的 KOL 身穿李宁 T 恤出现，从漫威到中国汽车品牌红旗等品牌都在排队。起初，李宁更像是一个“低成本”的品牌，顾客也不会炫耀，像迪斯尼这样的大牌也不会考虑联合品牌活动。

李宁管理团队决定瞄准富有的千禧一代，并定位为高端品牌。价格名列前茅，但这样消费者就会更愿意以同样的价格购买阿迪达斯或耐克。对于那些预算有限的人来说，其他本土品牌如特步（Xtep）或 361 度（361 Degrees）也能满足他们的需求，他们放弃了李宁品牌。对新口号“改变”的反应也不是很积极。以千禧一代为目标的新活动的结果是灾难性的，因为该品牌失去了其历史客户群，未能吸引到令人垂涎的群体。品牌累计亏损 30 亿元。

Surfing on the fact that global brands lost of bit of their shine and all brands tend to play it safe, Li Ning allocated considerable efforts to Research and Development and operated a phenomenal come back with the “Ways of Wade” collection featuring a winning combination of EVA midsole and rubber outsole with a carbon layer in between to absorb shocks and heel cushioning to bounce back.

Li-Ning co-created the “The Way of Wade” collection and customized the shoe to the NBA player needs so that he can proudly wear them during his games. The initials “ZZDG” and numbers 92 and 12116 on the sole stand for his sons Zaire and Zion, his nephew Dada, and girlfriend Gabrielle names and his grandmother’s, children and nephew ages. Wade admired Li Ning and they both connected on a shared vision for this special collaboration(Dwyer, 2021).

The recent “China Li-Ning” collection went sold out on Tmall, and KOL including Jolin Tsai and Bella Hadid appeared in a Li-Ning t-shirt and brands from Marvel to Chinese car brand Honq Qi are queueing. At first Li-Ning was more of a “low cost” brand and customers would not particularly show off about it and big names like Disney would not have considered co-branding activities.

Li-Ning management team decided to target wealthy millennials and position the brand as high end. Prices were on top but products not really so that consumers would prefer buying Adidas or Nike for the same price point. For those with a more limited budget, other local brands such as Xtep or 361 Degrees would do the trick and they turned their back on the Li-Ning brand. The response to the new catchline “make the change” was not very positive either. The results of the new campaign targeted at Millennials were disastrous as the brand lost its historical customer base and failed to attract the coveted group. The brand accumulated a total loss of 3 billion yuan.

2015 年，李宁重新夺回了领先地位，重新引入了其标志性的口号“一切皆有可能”：“任何一家公司在转型过程中都面临着巨大的挑战。人们习惯于在现有的轨道上前进。即使他们知道存在问题，第一反应就是要保护老办法。”“李宁公司近年来也遇到了巨大的挑战。但我不认为（错误的）责任在于经理和员工，而是在于规则。我们必须改变规则，以激励我们的人民不断改革，采取更准确和更有效的办法。然后他们可以和公司一起成长 (Wharton, 2016)。”

李宁决定实施五大战略转变：

- 孩子们穿。他发现了给孩子们穿衣服的机会，于是推出了李宁 Young 系列。

- 优化门店。门店应该传达良好的品牌体验，并围绕运动类别和生活方式进行组织。这项战略很难在小门店实施，因此被关闭了。

- 在线销售。网上购物不容忽视，特别是在中国大陆，人们不会在店内闲聊。为了采取更灵活的方法，生产周期和库存已经减少。电子商务迅速增长，占销售额的 20% 以上。

- 部落营销。针对特定群体，与说唱歌手 GAI 合作或赞助 3+1 街头篮球。

- 为运动员设计。研究和开发的努力是为了瘦身和减少汗水，推出了非常实用的跑鞋。

In 2015, Li Ning took back the lead and reintroduced its iconic catchline “anything is possible”: “Any company faces big challenges during transformation. People are accustomed to going ahead on the existing track. Even when they know there are problems, the first reaction is to protect the old way.” And “Li-Ning Co. has met huge challenges in recent years. But I don’t think the fault [for mistakes] lies with the managers and the staff; rather it is with the rules. We have to change the rules to motivate our people to constantly reform and adopt the more accurate and more efficient approaches. Then they can grow together with the company(Wharton, 2016).”

Li Ning decided to implement 5 major strategic changes:

-Children wear. Spotting the opportunity for dressing the little ones, he launched a Li Ning Young collection.

-Optimizing stores. Stores should convey a great brand experience and be organized around sports categories and lifestyle. The new strategy was difficult to implement in tiny stores which have consequently been closed.

-Selling online. Online shopping cannot be ignored, especially in mainland China where people are not fond of in-store chit-chat. Production lifecycle and inventory have been reduced for a more agile approach. E-commerce grew quickly to represent over 20% of the sales.

-Tribe marketing. Targeting specific groups partnering with rapper GAI or sponsoring 3+1 street basketball.

-Designing for athletes. Research and development efforts were geared towards thinness and sweat reduction, introducing very functional running shoes.

李宁与德维恩·韦德（Dwyane Wade）合作开发了篮球等传统项目，并资助了中国篮球队，流行的羽毛球和跑步。当李宁的核心品牌专注于专业运动和运动员时，该集团创建了户外运动艾高（Aigle）和红双喜乒乓球等垂直细分品牌。李宁一直在关注像匹克球这样的新兴运动 (Fellner, 2018)。

如果你是像我这样的新手，匹克球是网球、乒乓球和羽毛球的完美结合。你用一个木制的或复合材料的球拍和一个小球过网，通常是两人一组。这是一项对发球精度和速度要求较低的轻松运动，在养老院和游轮上非常受欢迎，在美国人、加拿大人和中国人中越来越流行。

随着加拿大华人社区的迅速发展，加拿大刚刚开设了以球拍为导向的展厅，以迎合当地该品牌的粉丝群。重点关注乒乓球、羽毛球和匹克球 (Toneguzzi, 2019)。

在阿迪达斯和耐克的较量中，时尚通过与斯特拉麦卡特尼（Stella McCartney）的合资企业或胶囊系列发挥了越来越大的作用。李宁把他五颜六色的运动服带上了 T 台，创造了惊喜。在纽约时装周的首次亮相，标志着这个现在被认为既实用又时尚的品牌迎来了一个转折点。

Li-Ning developed traditional categories like basketball with a notable collaboration with Dwyane Wade and the funding of the Chinese basket ball team, popular badminton and running. While the core Li Ning brand specializes in professional sports and athletes, the group created verticals like for Aigle outdoor sports or Double Happiness for table tennis. Li-Ning keeps an eye open on emerging sports like pickle-ball (Fellner, 2018).

If you are a novice like me, pickle-ball is a winning combination of tennis, table tennis, and badminton. You play it with a wooden or composite paddle and a small ball over a net, usually in two teams of two. Very popular in retirement houses and on cruise ships, the more relaxed sport that requires less service accuracy and speed, is growing over American but now also Canadian and Chinese people.

Racket oriented showrooms just opened in Canada to cater to the brand's local fan-base as the Chinese community is rapidly growing in Canada. Focus is on table-tennis, badminton, and pickle-ball (Toneguzzi, 2019).

In the battle between Adidas and Nike, fashion played an increasing role through joint ventures with Stella McCartney or capsule collections. Li Ning created the surprise by bringing his colorful sportswear on the runway. The debuts at New York fashion week marked a turning point in the brand that is now considered both functional and fashionable.

5.7 练习 Exercises

5A 趋势追踪

看看一个名人或设计师的社交媒体，他们不是在追随潮流，而是在创造一种独特的风格：他们风格的哪一方面会成为未来的时尚潮流？解释你的回答。

5B 智能手机的未来

想想智能手机的未来。为您喜爱的手机品牌制定创新路线图 (图 5.10) ：

问题 1：哪些与环境、生活方式、新用途、替代产品或政治相关的外部因素会对品牌未来 15 年的表现产生影响 ?

问题 2：对喜欢权力的客户来说，设备的三个关键特性是什么 ? 对喜欢性能的客户来说 ? 对喜欢人气的客户来说 ?

您可以在 DervalResearch 代戈微信上获得创新路线图框架 : 扫描二维码并键入关键词 **设计 5**。

5A Trend Hunting

Check the social media of a celebrity or designer that is not following trends but creating a unique style: what aspect of their style could become a fashion trend in the future? Explain your answer.

5B The Future of Smartphones

Consider the future of smartphones. Develop an innovation roadmap (Figure 5.10) for your favorite mobile phones brand:

Question 1. What external factors related to environment, lifestyle, new usages, substitution products, or politics could have an impact of the brand's performance coming 15 years?

Question 2. What would be the 3 key features of a device for a customer into power? for a customer into performance? for a customer into popularity?

You can get the innovation roadmap framework on DervalResearch 代戈 wechat: scan the QR code and type the keyword **design5**.

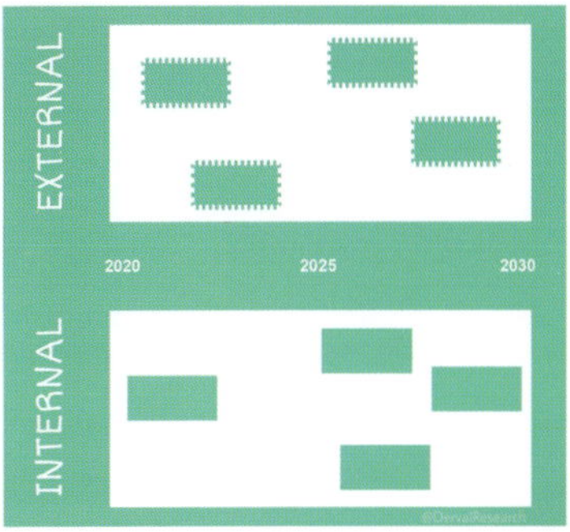

图 5.10 创新路线图框架 (经由 DervalResearch 代戈提供)
Figure 5.10 Innovation Roadmap framework (courtesy of DervalResearch 代戈)

5.8 要点 Take-Aways

- 观察和专家访谈是寻找趋势的有效方法
- 做一个榜样是开始潮流的方法，因为人们不想买像你一样的东西，他们想成为你
- 当人们给你的品牌或产品起绰号时，你就知道它是标志性的
- 许多生理因素都会影响购买决策
- 同样的人物角色，在同样的背景下，会以同样的方式表现

- Observation and expert interviews are an effective way to hunt trends
- Being a role model is the way to start trends, as people don't want to buy like you – they want to be you
- You know your brand or product is iconic when people create a nickname for it
- Many physiological factors can influence the purchasing decision
- The same persona, in the same context, will behave in the same way

5.9 术语表 Glossary

趋势

“文化规范深刻地反映了这一趋势，因此提出未来的趋势是可行的。” (Kim, 2007)

生物标志物

生物标志物（biomarkers）是“生物标记物”（biological markers）的缩写，用于“预测未来事件；相反，暴露的生物标志物用于确定过去的事件”。(Kuhnle, 2012)

激素

“激素是肽/蛋白质、脂类或氨基酸衍生物，分泌后作用于受体，在各自的感受性细胞内引起反应。身体的每一个细胞都以某种方式受到激素的调节。”

加伦（Galen）是皇帝和角斗士的医生，也是生理学之父，他已经注意到人们往往有不同的气质类型，如黏液质、多血质、胆汁质、抑郁质。他认为荷尔蒙的不平衡导致了这些性格特征。这些神秘的体液后来被发现是荷尔蒙，并产生了一个新的研究领域，称为内分泌学。（Brooks, 1962）

Trend

“Cultural codes deeply reflect on the trend and so that it is feasible to suggest the future trend.” (Kim, 2007)

Biomarkers

Biomarkers, contraction of “biological markers”, are used “to predict future events; conversely, biomarkers of exposure are used to determine past events”. (Kuhnle, 2012)

Hormones

“Hormones are peptides/proteins, lipids or amino acid derivatives which after their secretion act on receptors to cause reactions within the respective receptive cells. Each cell of the body is in one way or another regulated by hormones.”

Galen, physician to emperors and gladiators, and father of physiology, already noticed that people tended to have different temperaments – being phlegmatic, sanguine, melancholic, or choleric. He thought imbalances in humors was causing these personality traits. These mysterious humors were revealed much later to be hormones and gave birth to a new field of research called endocrinology. (Brooks, 1962)

5.10 参考文献 References

Derval, D. (2022). The Right Sensory Mix: Decoding Customers' Behavior and Preferences. Springer Nature, ISBN: 978-3-662-63794-4.

Derval, D. (2018). Designing Luxury Brands: The Science of Pleasing Customers' Senses. Springer Business & Management, ISBN: 978-3-319-71555-1.

Derval, D. (2018b, July 10). Qu'est-ce qu'un bon nom de marque ? Harvard Business Review, France. Retrieved from https://www.hbrfrance.fr/chroniques-experts/2018/07/21007-quest-ce-quun-bon-nom-de-marque/

戴安娜 · 代尔瓦勒 (Diana Derval). 奢侈品品牌营销：创建 · 实施 · 案例 (Luxury Brand Marketing). 东华大学出版社 (2016), ISBN: 9787566910424.

Derval, D., Bremer, J. (2012). Hormones, Talents, and Career: Unlock your Hormonal Quotient. SpringerBriefs, ISBN: 978-3-642-25712-4.

Derval, D. (2011). Réussir son étude de marché en 5 jours. Eyrolles, ISBN: 978-2-212-54958-4.

Lambert, F. (2020, March 31). Tesla's next killer app: Solar power on its electric cars — starting with Cybertruck. Electrek. Retrieved from https://electrek.co/2020/03/31/tesla-next-killer-app-solar-power-electric-car-cybertruck/

Kim, Y. K., & Lee, K. H. (2007). A Study on the Men's Fashion Trend through the Statistical Analysis. Journal of the Korean Society of Clothing and Textiles, 31(6), 837-847.

Kuhnle, G. G. (2012). Nutritional biomarkers for objective dietary assessment. Journal of the Science of Food and Agriculture, 92(6), 1145-1149.

Wuttke, W., Jarry, H., & Seidlova-Wuttke, D. (2010). Definition, classification and mechanism of action of endocrine disrupting chemicals. Hormones, 9(1), 9-15.

Brooks, C. M. (1962). Humors, Hormones, and Neurosecretions. SUNY Press.

Gertz, M. (2019, January 5) 171 Years Before Tesla: The Evolution of Electric Vehicles. Bloomberg. Retrieved from https://www.bloomberg.com/news/photo-essays/2019-01-05/171-years-before-tesla-the-evolution-of-electric-vehicles

Li-Ning (2021). The story of Li Ning. Li-Ning. Retrieved from http://en.lining.com/story.html

奶爸的球鞋潮流的喜好 . (2018, October 26). 国潮之光！李宁 x 红旗全新联名系列 .

L 圆圆 . (2019, July 12). 李宁的起死回生 . 亿欧 . Retrieved from https://www.iyiou.com/p/105260.html

Dwyer, R. (2021, January 8). Dwyane Wade and Li-Ning Debut the Way of Wade 9. Hyperbeast. Retrieved from https://hypebeast.com/2021/1/dwayne-wade-li-ning-way-of-wade-9-release-date-info

Toneguzzi, M. (2019, November 21). Chinese Sports Retailer 'Li Ning' Expands into Canada with 1st Storefronts. Retail Insider. Retrieved from https://www.retail-insider.com/retail-insider/2019/11/chinese-sports-retailer-li-ning-expands-into-canada-with-1st-storefronts

Fellner, D. (2018, April 1). Global-travel. Retrieved from https://global-travel-info.com/pickleball-cruising/

Wharton (2016, June 22). Lord of the Rings: Olympic Medalist Li Ning Rescues His Company. Knowledge@Wharton. Retrieved from https://knowledge.wharton.upenn.edu/article/lord-rings-olympic-gold-medalist-li-ning-returns-rescue-company-founded/

第六章 Chapter 6

设计生活方式品牌 Designing a Lifestyle Brand

琳琅满目 A feast for the eyes

在本章中，我们将评估生活方式品牌的持久影响。首先，我们想知道是什么让一些奢侈品和时尚配饰如此不可抗拒。如蒂芙尼（Tiffany&Co.）和施华洛世奇（Swarovski）案例所示，答案可能在于我们对闪亮的迷恋。我们将以万宝龙（Montblanc）为例，与理想的影响者合作。哈雷戴维森（Harley Davidson）和73小时（73Hours）将展示如何成为一个生活方式品牌。

目录

图 6.1 方方为 DervalResearch 代戈绘制的插画
Figure 6.1 Illustration by FangFang for DervalResearch 代戈

In this chapter we evaluate the long-lasting impact of lifestyle brands. First, we wonder what makes some luxury and fashion accessories so irresistible. As illustrated by the Tiffany & Co. and the Swarovski cases, the answer might reside in our fascination for shiny. We will make sure to collaborate with the ideal influencers following the Montblanc example. And Harley Davidson and 73Hours will showcase what it takes to become a lifestyle brand.

Content of this Chapter

6.1 为什么我们喜欢闪亮？
Why do we like Shiny?

我们先来个关于车和昆虫的小测验。根据你的说法，那些在湖边和其他池塘附近徘徊的水虫更喜欢：

（A）红色法拉利

（B）黄色兰博基尼

（C）白色特斯拉

（D）一辆黑色保时捷？

完成了？解决方案请查看第 116 页。

令人惊讶的是，即使是虫子在汽车方面也有自己的偏好。

光可是一个振荡的电磁场，以通过其偏振（或方向）、颜色和大小（或强度，如果您愿意的话）来表征。

当反射到黑色或红色发亮的表面时，比如一辆涂有油漆的汽车的车顶，光线的偏振度（或方向）很高且水平，使物体对水生昆虫非常有吸引力，可能会将其与水混淆。动物学家很长一段时间以来一直在想，为什么这么多水生昆虫在不撞上红色车顶的情况下成群结队地产卵。所以你知道在沼泽地里要避免哪种颜色来保持你的豪华车干净 (Derval, 2018) 。

图 6.2 汽车测试（经由 DervalResearch 代戈提供）
Figure 6.2 Car quiz (courtesy of DervalResearch 代戈)

Let's start by taking a quiz on cars and insects. According to you, would an aquatic bug – the ones hanging around near lakes and other water ponds – prefer:

(A) A red Ferrari

(B) A yellow Lamborghini

(C) A white Tesla

(D) A black Porsche?

Done? Check the solution page 116.

Surprisingly, even bugs have their preferences in terms of cars.

Light is an electromagnetic field that oscillates and can be characterized by its polarization (or direction), color, and magnitude (or intensity, if you prefer).

When reflecting on a black or a red shiny surface like the roof of a metal painted car, the polarization (or direction) of the light is high and horizontal making the object very attractive to aquatic bugs, confusing it with water. Zoologists had been for a long time wondering why so many aquatic bugs were swarming, laying eggs, when not crashing on the roof of red cars. So, you know which color to avoid to keep your fancy car clean in the swamp (Derval, 2018).

光线可以是：

–非偏振。就像蜡烛，没有特别的光芒方向。

–线性偏振（垂直或水平）。如汽车车顶或日落时。

–圆偏振。在自然界中很少出现，比如天空中的星星、海上的日落、螳螂虾和钻石（图 6.3）。

动物利用它们的磁感应和光的偏振来导航，寻找食物和饮料，与同伴交流，并寻找潜在的约会对象。

许多像索菲特这样的豪华酒店在室内设计中都考虑到风水（和磁场）。光偏振使得汽车和珠宝对一些人特别有吸引力。

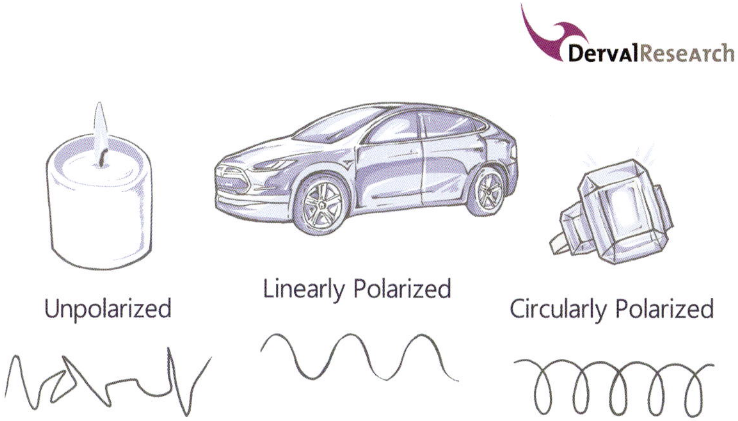

图 6.3 万宝龙影响者示意图（经由 DervalResearch 代戈提供）
Figure 6.3 Light Polarization (courtesy of DervalResearch 代戈)

Light can be:

–Unpolarized. Like in the case of a candle, beaming in no particular direction.

–Linearly polarized (vertically or horizontally), like in the case of the roof of a car or a sunset.

–Circularly polarized. More rarely present in nature, like in the case of stars in the sky, a sunset on the sea, mantis shrimps, and diamonds (Fig. 6.3).

Animals use their magnetic sense and the light polarization to navigate, find food and drinks, communicate with their peers, and get potential dates.

Many luxury hotels like Sofitel take feng shui (and the magnetic field) into consideration in their interior design. And the light polarization makes cars and jewelry particularly attractive to some.

6.2 蒂芙尼公司：将蓝色变成金色 Tiffany & Co.: Turning blue into gold

路易威登集团（LVMH）可谓是又一个适应顾客感知的成功案例，旗下品牌包括酩悦·轩尼诗（Moët & Chandon）、丝芙兰（Sephora）、路易威登（Louis Vuitton）、公主游艇（Princess Yacht）、迪奥（Dior）、高田贤三（Kenzo），最近还以 1000 亿日元收购了蒂芙尼（Tiffany）。令人垂涎的传奇品牌因其浪漫的珠宝而闻名，但大部分也因为其蒂芙尼蓝，又名潘通 1837——以成立的年份命名。在不知不觉中这个品牌利用了一种强大的光学现象将蓝色变成了金色：海丁格的画笔 (Derval, 2018)。

我们的黄斑位于视网膜中央，它的工作原理类似于一个径向分析仪，特别是滤过蓝光，峰值为 460 纳米。它似乎是通过把蓝色变成金色来过滤的，创造了一种叫做海丁格的画笔的现象（图 6.4）。这种现象最好是在蓝天日出或日落的天顶上用主眼观察（Horváth, 2014）。当被这些蓝色所淹没时，视网膜中的一个系统正在把它的一部分变成金色，这样你只能看到某种金色蝴蝶结周围的蓝色。

蒂芙尼（Tiffany）的旗舰店在正面展示中使用了浪漫的“蓝盒子”标志品牌。橱窗里的陈列物实际上被辐射蓝光包围着——一种激活我们磁感受器的光。

图 6.4 海丁格的画笔（经由 DervalResearch 代戈提供）
Figure 6.4 Haidinger's Brush (courtesy of DervalResearch 代戈)

LVMH – Louis Vuitton and Moët Hennessy luxury group – is another great example of successfully adapting to local perception with brands like Moët & Chandon, Sephora, Louis Vuitton, Princess Yacht, Dior, Kenzo, and recently acquired Tiffany for 100 billion ¥. The coveted legendary brand got famous for its romantic jewelry but mostly also for its Tiffany Blue, aka Pantone 1837 – named after the year of foundation. Unknowingly, the brand harnessed a powerful optical phenomenon turning blue into gold: the Haidinger's Brush (Derval, 2018).

Our macula – an area located at the center of the retina – works like a radial analyzer filtering especially the blue light, with a peak at 460 nanometers. It seems that it is filtering by turning blue into gold, creating a phenomenon called the Haidinger's Brush (Fig. 6.4). This phenomenon is best observed at the zenith of a blue sky sunrise or sunset and with the dominant eye (Horváth, 2014). When submerged by all this blue, a system in the retina is turning part of it into gold so that you only see the blue around some kind of golden bow-tie.

Tiffany's flagship stores use the romantic "blue box" signature branding in the front display. The window display is literally surrounded by radiant blue light – a light activating our magnetoreceptors.

难怪感偏振光女性（以及男性）会像磁铁一样被这些物品所吸引。在海丁格的画笔效应下，一些女性看到一半的盒子是蓝色的而另一半是金色的，也是不例外的！——甚至更好！因此，蒂芙尼（Tiffany）不需要用黄金来打动奢侈品爱好者，因为黄金在购物者的眼中。

蒂芙尼还以其钻石公主切割而闻名。刻面越多，钻石就越有光泽，因为光线反射到更多的侧面。最近的 T 系列以更现代的纽约为灵感，提出了模仿 Tiffany T 的简洁线条。这一次，除了历史悠久的纯银外，大胆的单品还采用了各种细微的黄金（白色、黄色和玫瑰色），以增添额外的光泽。

图 6.5 “如果蒂芙尼会做鞋”，概念由 Aubery 和 Jessie 提供，
Aubery 为 DervalResearch 代戈绘制的插画
Figure 6.5 "If Tiffany would make a shoe", concept by Aubery and Jessie,
Illustration by Aubery for DervalResearch 代戈

No wonder polarotactic women (and men) are attracted to the display like magnets. It is not to be excluded even that some women, under the Haidinger's Brush phenomenon, see half of the box blue and the other half gold – even better! So, Tiffany does not need to use gold to impress luxury lovers as the gold is in the eyes of the shoppers.

Tiffany is also famous for its diamond princess cut. The more facets you have, the shinier is the diamond as the light reflects on more sides. The recent T collection, inspired by a more modern New York, proposes clean and simple lines mimicking the T from Tiffany. This time, in addition to the historical sterling silver, the bold pieces come in all nuances of gold (white, yellow, and rose) for an extra shine.

6.3 影响者地图：万宝龙（Montblanc）案例 Influencers' Map : the Montblanc case

古人用剑比试武艺高下，之后商人用钢笔谈判，现在土豪用金色苹果手机 iPhone 彰显身份，而明天你又会带什么去会议室让商业伙伴眼前一亮呢——也许是一块智能手表？历峰集团（Richemont group，也以卡地亚 Cartier、克洛伊 Chloé 和积家 Jaeger-LeCoultre 等品牌闻名）旗下的万宝龙（Montblanc）了解这种转变，现在把重点放在手表以及从手提箱到耳机等其他装备上，作为其品牌战略。关键意见领袖和其他有影响力的人在万宝龙发挥着重要作用(Derval, 2016)。

运用影响者地图的目的是明确每个目标群体在购买时会以谁为榜样。他们会追随演员？还是政客？一般那些敢冒生命危险的人是男性的榜样：登山运动员、水手、极限运动员、宇航员、飞行员，当然还有超级英雄。难怪万宝龙（Montblanc）会选择《复仇者联盟》中雷神索尔的扮演者克里斯·海姆斯沃斯来推广他们的产品了。品牌其实还任命了另外一名超级英雄，即《X 战警》中金刚狼的扮演者休·杰克曼为其全球代言人。

图 6.6 方方为 DervalResearch 代戈绘制的插画
Figure 6.6 Illustrations by FangFang for DervalResearch 代戈

In ancient times, deals were negotiated with the sword, then with fountain pens, more recently with a golden iPhone, and tomorrow with the next device you can bring to the meeting room to impress your business contacts – maybe a smart watch? A luxury pen maker like Montblanc, part of the Richemont group (also known for their brands Cartier, Chloé, and Jaeger-LeCoultre), understood this shift and is now focusing on watches, and other gear, from suitcases to headphones, in its branding strategy. Key Opinion Leaders and other influencers play an important role at Montblanc(Derval, 2016).

The idea of the Influencers' map is to identify for each target persona who they look up to. Do they admire and follow actors? Politicians? Men tend to take on as role models those who risk their lives: mountain climbers, sailors, extreme sportsmen, astronauts, pilots, and of course super heroes. No wonder Montblanc selected Chris Hemsworth, the actor playing the role of Thor, God of thunder, in the Avengers movie, to promote their products. The brand even appointed another super hero, Hugh Jackman, aka Wolverine in X-Men, to be its global brand ambassador.

万宝龙（Montblanc）肯定了杰克曼形象与品牌一致并说明“他代表了品牌的所有属性：优雅，能干，有开拓性，敬业”。许多奢华手表制造商会找高尔夫运动员和F1赛车手为目标代言人，但是若你发现你的用户其实喜欢打羽毛球，该怎么办呢？万宝龙（Montblanc）就任命了中国羽毛球运动员林丹为其大中华区的代言人。林丹被他的同辈称作“兼具风格和本事的男人”。当你的用户喜欢骑马，并暗自将一切同皇室关联起来时，夏洛特·卡西拉吉是品牌理想的代言人。需要注意的是万宝龙（Montblanc）任命的品牌代言人夏洛特不仅会骑马，还是时尚偶像和摩纳哥的公主！让有着百年传统的生产 Meisterstück 高端产品的这个品牌呈现出拟人化的形象，唯有冠军才能担此重任 (Derval, 2016)。

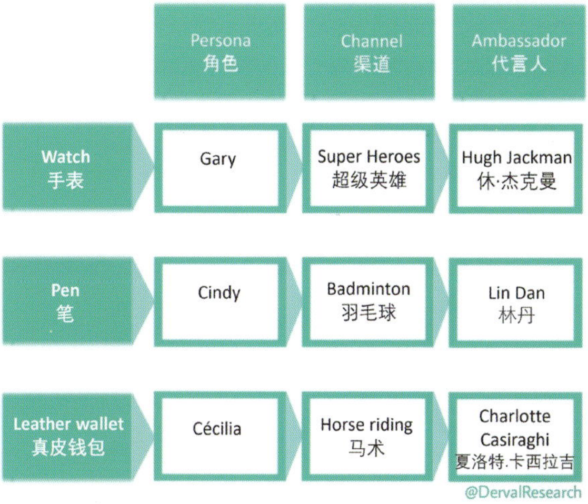

图 6.7 万宝龙影响者地图（经由 DervalResearch 代戈提供）
Figure 6.7 Montblanc Influencers Map (courtesy of DervalResearch 代戈)

The brand confirms the congruence by stating that Jackman “represents all the attributes of the Montblanc brand: elegant, talented, pioneering, and committed to the arts”. Many luxury watch manufacturers target golf players and Formula 1 racers but what if you discover that your personas actually fancy badminton? Then like Montblanc you appoint the Chinese badminton player Lin Dan as your brand ambassador. Lin Dan is recognized by his peers as a “man of style and substance”. When your personas fancy horse-riding and also secretly everything related to royalty, Charlotte Casiraghi is the ideal brand ambassador. Note that this eminent horse jumper appointed as brand ambassador by Montblanc also happens to be a fashion icon and Princess of Monaco! A task for champions only as it is difficult to impersonate the brand’s 100-year heritage with sublime pieces like the fountain pen Meisterstück (Derval, 2016).

6.4 73 小时：从头到脚
73 Hours: from Head to Toe

短短几年，高跟鞋 73 小时已成为上海乃至全国的热门鞋履品牌，从成都到北京都有精品店。该品牌以中国女性白领为目标，以 1000 元左右的合理预算提供一流的工艺。这个角色——不反对网上购物——非常明确，可以有特定的口号，比如“只做时髦且耐磨的高跟鞋”。来自 73 小时团队的肖倩证实，该品牌想为现代女性提供这样一种一站式购物体验，“她们可以在店里剪头发、做美甲、买鞋，还可以在咖啡厅和朋友一起喝茶或喝咖啡。这就是我们在香港广场商场开设上海旗舰店，并邀请合作伙伴加盟的原因。到目前为止市场反应非常好。”

该品牌在通过社交媒体（淘宝，微博，微信，天猫）分享高跟鞋穿着技巧的一众鞋迷中找到了自己的核心角色。因此，在有限的营销预算下，73 小时能够吸引许多人成为自己的忠实用户。他们的秘诀就是每周尽最大努力用新款式来满足鞋迷的需求。将快时尚的原则应用到高端鞋业并不容易，因为它需要缩短生产周期，以满足客户的期望——要知道 73 小时社交媒体上最常见的关键词是“新产品”。目标客户是在工作场所担任高级职务的妇女。该品牌还明确每双鞋从打磨到送至顾客手中至少需要 73.3 小时的时间。对于那些不仅看重国际品牌美誉度，而且欣赏高端产品的人物来说，这是非常令人满意的。

Within just a couple of years, 73Hours has become a hot shoe brand in Shanghai and across the country with boutiques from Chengdu to Beijing. The brand targeted at Chinese female white-collars offers top notch craftsmanship for a reasonable budget of around 1000 yuan. The persona – not opposed to online shopping – is super clear, which allows specific catchlines such as “high-heeled shoes that are fashionable as well as wearable”. Xiao Qian, from 73Hours team, confirms that the brand wanted to propose modern women a full experience “they can get a haircut, have their nails done, buy shoes in the store, but also enjoy a tea or a coffee break with friends in the café. This is the reason why we set up Shanghai flagship store in HK Plaza mall and invited partners to join. So far the market response has been really good”.

The brand found its core persona among shoe maniacs who share tips and recommendations via social media (Taobao, Weibo, Wechat, and Tmall). So that. with a limited marketing budget, 73Hours was able to convert many new clients. Their secret is that they do their best to address the needs of shoe fans with new styles every week. Applying fast fashion principles to the high-end shoe industry is not easy, because it requires to shorten the production cycle in order to meet customers’ expectations – knowing that the most common keyword on 73Hours social media is “new product”. The target clients are women assuming higher functions in the workplace. The brand name also clearly emphasizes that it takes at least 73.3 hours to assemble, polish and deliver each pair of shoes to the customers. For these personas, who do not just value the reputation of international brands, but also appreciate high-end products, this is very satisfying to know.

鞋子甚至被装在一个专为装珠宝而设计的盒子里，盒子里有一个鞋形的切割泡沫垫，这样鞋子才能被妥善保管并安全运输。肖倩强调，73 小时的鞋子提供了一种独特的“优雅，戏剧性和女性气质”的组合。

凭借精美的设计、103 道工序和高品质的羊皮，该品牌为“都市女性提供时尚且舒适的鞋子”(图 6.8)。简单的定位，但在 73 小时品牌成立之前却几乎很少有品牌真正实现过。肖倩透露，“73 小时特别研发的鞋子非常适合中国人和亚洲人的脚，鞋跟更低，前脚更宽”(图 6.8)。73 小时市场总监表示，未来的重点将是在中国开设更多的门店，推出新的鞋款、个性化服务和更严格的供应链管理。

图 6.8 73 小时的情人节活动（经由 73Hours 提供）
Figure 6.8 73 Hours Valentine's Day campaign (courtesy of 73Hours)

73Hours shoes are even sold in a box designed to contain jewelry, with a shoe-shaped cut foam pad, so that shoes can be properly secured and transported safely. Xiao Qian stressed out that the 73Hours shoes offered a unique combination of “elegance, drama, and femininity”.

With the help of fancy designs, 103 procedures, and high quality sheepskin the brand provides “fashionable yet comfortable shoes for urban women”(Figure 6.8). The positioning is simple, but has never been achieved so far. “73Hours”, reveals Xiao Qian, “puts special research and development efforts into perfectly fitting the shoes to the Chinese and Asian feet, with a lower heel and a wider forefoot” (Figure 6.8). 73Hours marketing director shares that the focus for the coming time will be on opening more stores in China, coming up with new shoe designs, personalized services and a tighter supply chain management.

6.5 发展像施华洛世奇这样的家族企业 Developing a family business like Swarovski

这个奥地利创意品牌成功地成为了全球最受消费者和名人追捧的水晶品牌。最初它只是生产小件水晶雕像的品牌，后来逐渐开发了时尚配饰、珠宝、甚至可穿戴饰品。Joan Ng 为品牌转型做出了重大贡献，确保品牌适应本土顾客的感知。我们有幸采访了她，让她同我们分享一些真知灼见。

施华洛世奇执行董事会中国顾问、Bang&Olufsen 中国咨询委员会成员 Joan Ng 分享了 5 个成功的关键 (Derval, 2016)。

关键 1：定位和视野

Joan 在施华洛世奇工作 20 年了，一直积极地推动品牌从一个水晶雕像品牌转型为一个珠宝时尚品牌："不同于大多数面向精英群体的奢侈品品牌，施华洛世奇（Swarovski）的理念是为消费者提供买得起的优质产品。施华洛世奇（Swarovski）不是小众品牌。我们希望为所有渴望绽放光芒的女性带来璀璨夺目的魅力。"

关键 2：适应当地的经验

图 6.9 方方为 DervalResearch 代戈绘制的插画
Figure 6.9 Illustration by FangFang for DervalResearch 代戈

The innovative Austrian brand Swarovski managed to become the most sought-after sparkle brand for consumers and celebrities around the world. It all started with small crystal figurines and developed into fashion, jewelry, and even wearables. Joan Ng has been contributing to this shift and making sure the brand adapts to local perception. She accepted to share some insights.

Joan Ng, China Advisor to the Executive Board at Swarovski, and Member of the China Advisory Committee at Bang & Olufsen, shared 5 keys to success (Derval, 2016).

Key 1: Positioning and vision

Joan has been with Swarovski for 20 years and took an active part in shifting it from a crystal figurines brand into a jewelry fashion brand: "Unlike most luxury brands, who target an elite, Swarovski's philosophy is to propose the best products at an acceptable price. Swarovski does not want to exclude anyone. We want to give sparkle to any woman who wants a wonderful shine."

Key 2: Adapting the experience locally

施华洛世奇团队监测趋势，并为印度、日本、韩国和中国推出特别系列。Joan 强调说：“中国对高级珠宝的需求很大。销售渠道方面，我们有自营店也有合作伙伴，合作伙伴指获得我们许可的并能遵守品牌严格规定的零售店。我们的理念是让顾客无论身处哪家店均能感受到统一的建筑风格、视觉陈列和品牌体验。”施华洛世奇的流行产品在全球 2300 家商店和 8000 家多品牌商店有售。

关键 3：使客户为之叫绝

“毫无疑问顾客喜爱施华洛世奇（Swarovski）产品闪闪发光的特点，它的光芒来自于准确的切割，完美的水晶和打磨。我们的工厂拥有能工巧匠，经他们手工打磨之后，无论从哪个角度看，首饰都将闪烁最璀璨的光泽。注重细节是品牌的宗旨，产品完美无瑕的质量是我们骄傲的源泉。施华洛世奇（Swarovski）产品蕴含铑——一种非常昂贵的元素，它能更有效地防止产品磨损。”

关键 4：发明和创新

“施华洛世奇发明了如何将水晶切割到完美的程度，从那时起，就从未停止过创新”，Joan 回忆道，她是健康和美容技术领域的先驱，推出了活动追踪珠宝系列“活动水晶”。基于对女性健康和时尚意识的观察，“施华洛世奇设计了一个闪闪发光的时尚配饰系列，可以帮助监测她们的活动和睡眠，别具风格”。

关键 5：直接面向消费者

Joan 强调，如果一个家族企业的成功能延续到第五代，“品牌一定贯彻了某种正确的商业模式”。她认为施华洛世奇（Swarovski）致力于同我们的消费者建立联系。我们能在零售店内直接聆听消费者，不断适应他们的需求。

Swarovski teams monitor trends and released special collections for India, and for Japan, Korea, and China. Joan highlights: “ Fine jewelry is in demand in China. Regarding distribution, we have our own stores as well as partners – retailers who are allowed to sell our products provided they comply with the brand’s strict rules. The idea is that in terms of architecture, visual merchandising, and brand experience, customers cannot tell the difference.” Swarovski popular products are available in 2300 shops and 8000 multibrand stores worldwide.

Key 3: Wowing customers

“What customers love in Swarovski is without a doubt the sparkle. Sparkle comes from the precision-cut crystal and the polish. Our skilled workers polish each jewelry piece by hand to give each angle a maximum of shine.” reveals Joan, “our brand always paid attention to details and takes pride in having everything checked and checked again. We finish the products with Rhodium – a very expensive and sparkly component that will better protect against abrasion.”

Key 4: Inventing and Innovating

“Swarovski invented how to cut crystal to its perfection and, since then, never stopped innovating” reminds Joan who pioneered the field of health and beauty-tech with the activity tracing jewelry collection “Activity Crystal”. Based on the observation that women are both health and fashion conscious, “Swarovski designed a sparkly fashion accessory collection that can help monitor their activity and sleep, in style”.

Key 5: Direct to Consumer

Joan highlights that if a family business is still successful in the 5th generation “they must be doing something right” and explains that Swarovski is connecting with its consumers thanks to their retail stores, so that the brand can listen to them directly and continuously adapt to their demands.

6.6 哈雷戴维森（Harley Davidson）：摩托车 Harley Davidson: the bike

哈雷戴维森（Harley Davidson）以其独特的声音和酷炫的低座椅成为了一个生活方式品牌。

摩托车别出心裁的设计使其成为唯一可以让女性和“高度受到挑战的人”双脚落地的机车，因此在这两类人中十分受欢迎——这对驾驶摩托也十分有益，否则你的第一个转弯也许就是人生中最后一个了。

哈雷戴维森（Harley Davidson）显然明白他们的主打业务不是从 A 地骑行到 B 地，而是休闲骑行。自 1905 年成立以来，品牌就一直同摩托车赛事有关，它的每个广告都像是一则短小精悍的故事，娓娓道来哈雷戴维森（Harley Davidson）是如何让你的骑行之旅变得刺激和精彩。哈雷戴维森（Harley Davidson）传播的价值观与红牛（Red Bull）类似，都与冒险和寻求刺激有关 。重型摩托车市场包括传统摩托车、运动摩托车和两轮公路越野摩托车，米沃奇（Milwaukee）主打巡航和旅行摩托车，这些细分群体注重时尚款、定制款和例如能放置行李箱或挂包的旅行款 (Derval, 2016) 。

图 6.10 方方为 DervalResearch 代戈绘制的插画
Figure 6.10 Illustration by FangFang for DervalResearch 代戈

Harley Davidson became a lifestyle brand with its distinct sound and cool low seating.
The particular design of the motorbike made it very popular among women and “vertically challenged people” as it’s the only big bike that enables their feet to touch the ground – very useful to pilot a motorbike because otherwise the first turn might be the last.
Harley Davidson clearly understood their business was not about motorcycling from A to B but about leisure riding. Since its inception in 1905, the brand was associated with motorbike racing and each advertisement was like a mini story, telling how Harley Davidson makes your journey fierce and awesome. Harley Davidson was conveying values similar to Red Bull at that time and linked to recklessness and sensation seeking. Within the heavyweight motorbikes market, that includes traditional bikes, sport bikes, and dual bikes for off-road use, the Milwaukee brand is playing in the field of cruiser and touring bikes. Those sub-segments focus more on styling, customization, and traveling, with for instance large luggage compartments or saddlebags (Derval, 2016).

哈雷戴维森（Harley Davidson）善于建立分销渠道关系，原因在于控制分销渠道是跨国销售的关键。员工培训在传授经验方面亦发挥着巨大作用。为了避免“分销陷阱”效应，哈雷戴维森（Harley Davidson）注重与哈雷车主会（HOGs）直接联系，建立强大的社群。

哈雷戴维森（Harley Davidson）是一个提倡独特经历和增强体验的主打生活方式的品牌。它适应当地客户和建立强大社区的能力在印度和其他国家中发挥了重要作用。摩托车销售占品牌收入的四分之三，相关零部件和配件，包括摩托车服装系列在内的商品以及从拉斯维加斯芯片到打火机的许可证都增加了摩托车销售额。毫无疑问的是，哈雷戴维森（Harley Davidson）擅长传播独有的信息——即使是刺耳的噪音——来讲述品牌故事，同时还善于同车主会建立联系。他们的下一步就是顺理成章地为顾客提供全方位的生活方式体验。同宝马（BMW）推出信用卡的做法类似，哈雷为骑手推出了附带奖励机制的 Visa 卡。

这些相关服务不断地保持着品牌与顾客的联系，加强了顾客在社群中的归属感。这是一个多么庞大的社群：有 100 多万名车主加入哈雷车主会（HOG）。通过品牌安排的赛事、音乐会和其他派对活动，车主会积极地运作着。社群营销十分有效。

哈雷现在希望通过向注重环保的车手开放来扩大其客户群：哈雷推出了其首款电动摩托车 LiveWire，并正在推出 Serial 1。这款以 1903 年发布的第一辆哈雷命名的令人惊叹的电动摩托车线条简洁，但设计却很强大，有着独特的品牌代码，比如白色轮胎，让人想起了传说中的 Serial Number One(Hawkins, 2020)。

Harley Davidson invested in the relationship with the distribution channels, as it is key to keep the control over it when going abroad. Training the staff also played also a huge role in conveying the experience. To avoid a “distribution trap” effect, Harley Davidson took care of connecting directly with their Harley Owners Group also called HOGs, building a strong community.
Harley Davidson is a lifestyle brand proposing a unique and augmented experience. Its ability to adapt to local customers and build a strong community was instrumental in India, among other countries. The motorcycles sales that represent three fourth of the brand revenues are augmented by the related parts & accessories, merchandise including the MotorClothes line, and licensing from Las Vegas chips to lighters. There is no doubt that Harley Davidson is good at delivering a unique message – even a distinct noise – telling a brand story, and connecting with its HOGs. The next step is naturally to provide the full lifestyle experience. Similarly to what BMW is doing launching a credit card, Harley Davidson proposes a Harley Davidson Visa card, with a reward system for riders.
These associated services help keep a constant connection with customers and increase the feeling of being part of the community. And what a community: Harley Owners Group (HOG) counts over 1 million riders. The brand keeps the community active by arranging races, music concerts, and other parties. Community marketing works.
The brand now wants to broaden its customer base by opening it to environment-conscious riders: Harley launched the LiveWire, its first electric motorbike, and is now introducing Serial 1. The stunning electric bike, named after the first Harley released in 1903, has clean lines yet a powerful design with distinct brand codes like the white tires, reminiscent of that legendary Serial Number One (Hawkins, 2020).

6.7 练习 Exercises

6A 名人代言

思考一下你最爱的明星，他 / 她代言了什么品牌？在什么情况下代言？与哪种爱好有关？或者他们会出席什么类型的活动？谁会追随他们？为什么？可以用自身的例子解释。

6B 蒂芙尼大使

阅读 6.2 中 Tiffany & Co. 的案例，观察它们的产品，思考他们的目标用户角色模型，为品牌选择理想的代言人明星，说明原因。

您可以在 DervalResearch 代戈微信上获得影响者地图模板 (图 6.11): 扫描二维码并键入关键词 **设计 6**。

汽车测试的答案

正确答案是（D），然后是（A）。

6A Celebrity Endorsement

Think of your favorite celebrity: what brand is he/she endorsing. In which context? While practicing which hobbies? Or showing up at what type of event? Who are the people following him/her and why? You can use your personal example to explain.

6B Tiffany Ambassador

Read section 6.2 on Tiffany & Co., check their assortment and select 3 products. For each product, think of the target persona, the social media or hobbies they practice, and suggest a celebrity that would be the ideal brand ambassador. Explain why.

You can get the Influencers' Map template (Figure 6.11) on DervalResearch 代戈 wechat: scan the QRcode and type the keyword design6.

Solution to the cars quiz

The correct answer is (D) followed by (A).

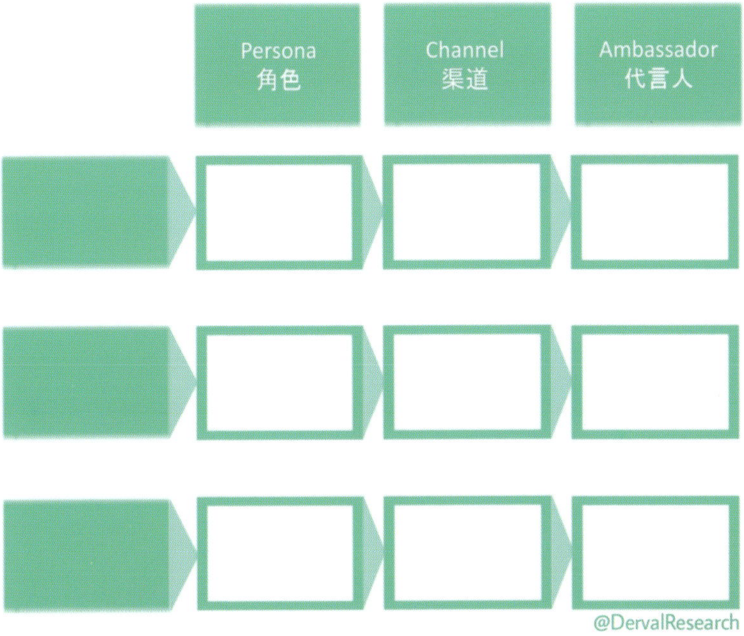

图 6.11 影响者地图模板（经由 DervalResearch 代戈提供）
Figure 6.11 Influencers' Map Template (courtesy of DervalResearch 代戈)

6.8 要点 Take-Aways

- 打造一个生活方式品牌可以增加销售和交叉销售的机会
- 只有生活方式品牌才有助于与目标人物角色进行深层次的沟通
- 与有影响力的人和品牌大使合作可以增加亲和力，但前提是要根据目标人物角色的愿望和兴趣对他们进行良好筛选
- 共同的活动和价值观是关键
- 即使是特殊的联系也无法弥补产品和体验的不足，一个生活方式品牌需要两者兼而有之

- Building a lifestyle brand multiplies the upselling and cross-selling opportunities
- Only a lifestyle brand can help deeply connect with the target personas
- Partnering with influencers and brand ambassadors can increase the affinity, but only if they are well screened, based on the target personas aspirations and interests
- Bonding over shared activities and values is the key
- Even an exceptional connection cannot make-up for poor products and experiences – a lifestyle brand needs both

6.9 术语表 Glossary

生活方式品牌

“生活方式品牌是一家公司，它销售其产品或服务，以体现一个群体或文化的利益、态度和观点。”(AMA, 2021)

磁感应

“许多动物依靠地球磁场进行空间定位和导航。”(Wu et al., 2012)

偏振光

“光是由随机振动的电磁波组成的。当这种振动被限制在一个特定的方向或以特定的方式被限制时，光被称为偏振光。”(Ito et al., 2019)

Lifestyle brand

“A lifestyle brand is a company that markets its products or services to embody the interests, attitudes, and opinions of a group or a culture.” (AMA, 2021)

Magnetic sense

“Many animals rely on Earth’s magnetic field for spatial orientation and navigation.” (Wu et al., 2012)

Polarized light

“Light consists of electromagnetic waves vibrating in random directions. When such vibrations are limited to one specific direction or limited in a particular way, the light is called polarized.” (Ito et al., 2019)

6.10 参考文献 References

Derval, D. (2022). The Right Sensory Mix: Decoding Customers' Behavior and Preferences. Springer Nature, ISBN: 978-3-662-63794-4.

Derval, D. (2018). Designing Luxury Brands: The Science of Pleasing Customers' Senses. Springer Business & Management, ISBN: 978-3-319-71555-1.

戴安娜 · 代尔瓦勒 (Diana Derval). 奢侈品品牌营销：创建 · 实施 · 案例 (Luxury Brand Marketing). 东华大学出版社 (2016), ISBN: 9787566910424.

Derval, D., Bremer, J. (2012). Hormones, Talents, and Career: Unlock your Hormonal Quotient. SpringerBriefs, ISBN: 978-3-642-25712-4.

DervalResearch (2020). The science of colors: From product to experience design. Iversity, Springer Nature Group. https://iversity.org/en/courses/the-science-of-colors

Horváth, G., Takács, P., Kretzer, B., Szilasi, S., Száz, D., Farkas, A., & Barta, A. (2017). Celestial polarization patterns sufficient for Viking navigation with the naked eye: detectability of Haidinger's brushes on the sky versus meteorological conditions. Royal Society open science, 4(2), 160688.

Hawkins, A. J. (2020, October 27). Harley-Davidson unveils a gorgeous new electric bike called serial 1. The Verge. https://www.theverge.com/2020/10/27/21536726/harley-davidson-serial-1-electric-bike-ebike-specs-price

AMA, 2021. American Marketing Association. Retrieved from https://www.ama.org/

Wu, L. Q., & Dickman, J. D. (2012). Neural correlates of a magnetic sense. science, 336(6084), 1054-1057.

Itoh, Y., Langlotz, T., Iwai, D., Kiyokawa, K., & Amano, T. (2019). Light attenuation display: subtractive see-through near-eye display via spatial color filtering. IEEE Transactions on Visualization and Computer Graphics, 25(5), 1951-1960.

第七章 Chapter 7

实现完美心流 Achieving the Perfect Flow

喜形于色 "Feeling a glow of happiness"

在这一章中，我们将研究如何实现完美的心流。熊、猴子和香蕉测试将提示如何适应不同大脑结构和期望值的客户。土豪金 iPhone 和奈斯派索(Nespresso)的案例将展示找到正确定位和识别真正竞争对手的重要性。完美的心流对在线、店内以及室内和景观设计都至关重要，我们将在酒店挑战赛中看到这一点。我们将详细研究戴森（Dyson）和古驰（Gucci）等品牌是如何持续创新的。

目录

图 7.1 Aubery 为 DervalResearch 代戈绘制的插画

Figure 7.1 Illustration by Aubery for DervalResearch 代戈

In this chapter, we will study how to achieve the perfect flow. The bear, the monkey, and the banana test will give hints on how to adapt to customers with different brain structures and expectations. The golden iPhone and Nespresso cases will demonstrate the importance of finding the right positioning and identifying the real competitors. The perfect flow is critical online, in-store, but also in interior and landscape design as we will see with the Hotel Designs challenge. And we will examine in detail how brands like Dyson and Gucci are able to continuously innovate.

Content of this Chapter

7.1 熊、猴子、香蕉 The Bear, the Monkey, and the Banana

这里有一只熊，一只猴子，一根香蕉：你怎么把它们分成两组？

通过这个简单的测试，可以了解更多关于大脑中数据的组织方式以及如何帮助您管理创新和设计的信息。完成了吗？结果请参见第 138 页。

我们在大脑中整理数据的方式取决于多巴胺 D2 受体的数量和分布。因此，当我们开始设计时，问题在于我们应该如何为目标受众安排形状和空间。大脑结构非常有条理的人会期望设计具有清晰的类别，而大脑就像思维导图的人会享受更加整体的设计 (DervalResearch, 2020)。

图 7.2 一只熊，一只猴子和一个香蕉
（图画经由 DervalResearch 代戈提供，作者：Vlad Kolarov）
Figure 7.2 The Bear, the Monkey, and the Banana
(Drawing courtesy of DervalResearch 代戈 , artist: Vlad Kolarov)

Here are a bear, a monkey, and a banana: how would you split them into two groups?
With this simple test, learn more on how data is organized in the brain, and how it can help you manage innovation and design. Done? See the results page 138.
The way we organize data in our brain depends on the number and the distribution of dopamine D2 receptors. So, when we start designing, the question is how should we arrange shapes and spaces for our target audience. People who have a very structured brain will expect a design with clear categories, while people who have a brain like a mindmap will enjoy a more holistic design (DervalResearch, 2020).

7.2 替代产品：金色苹果手机（iPhone）案例
Substitution products: the Golden iPhone case

苹果手机（iPhone）仅在中国的销售额就已超过 3000 亿人民币，土豪金迅速售罄。是什么类型的顾客会花 4500 元人民币买土豪金？最重要的是，为什么呢？某些中国百万富翁因“土豪”身份而出名。这类新贵一族靠采矿业或建筑行业发财致富后和家人们过上了奢侈的生活。DervalResearch 代戈以时装零售店员和专家的采访为基础发布了土豪的典型形象：身材不高，体型略发福，平均单次服装花费 30000 元人民币，总是忙着在电话上谈生意，出差时喜欢带着“兄弟”购物，追求昂贵的服装。衣服上再多的金色或再大的标志对他们而言也不为过。路易威登（LV），爱马仕（Hermès）和古奇（Gucci）等奢侈品领导品牌最受他们青睐。

土豪这么做是为了引起谁的注意？女性？不，那样他的信用卡就够了 (Derval, 2016)。土豪是要震慑其他男性，而且这确实很有效！

所以土豪不戴手表而是大的手表，不开汽车而是大的汽车。试想猫或青蛙为了吓退对手会让自己看上去更壮硕，我们也就不难理解土豪为什么要这么做了。昨天参加会议，展示万宝龙笔，今天展示金色 iPhone，明天展示欧拉戒指，一定会给观众留下深刻印象，对谈判有帮助。记住，中国古代有权有势的人就经常佩戴黄金或玉器，如今金色成为继房产和汽车之后中国有钱人的新宠也就不足为奇了。

图 7.3 方方为 DervalResearch 代戈绘制的插画
Figure 7.3 Illustration by FangFang for DervalResearch 代戈

The iPhone generated over 300 billion rmb revenues just in China and the golden edition quickly went sold out. But what type of customer would put 4500 rmb in a golden phone and most importantly why? Some Chinese millionaires are also known as Tuhao. These new rich made their fortune in the mining or construction industry and are now enjoying with their family a luxury lifestyle. Based on interviews with fashion retail vendors and experts, DervalResearch 代戈 released the portrait of a typical Tuhao. His average clothing basket would be 30 000 rmb. Short, a little belly, always busy on the phone for business, he likes to shop with “his bros” while on a business trip, and is looking for expensive clothing. He can never wear too much gold or a too big brand logo. His favorite brands are the leaders in their field: Louis Vuitton, Hermès, and Gucci.

Who is the tuhao trying to impress: women? No, for that his credit card is convincing enough (Derval, 2016). The tuhao is trying to impress other men. And it works!

That is why he is not wearing a watch, but a big watch, and he is not driving a car, but a big car. Think of animals like cats or frogs trying to make themselves look bigger than they are in order to scare away competitors. Joining a meeting and exhibiting a Montblanc pen yesterday, a golden iPhone today, and an ŌURA ring tomorrow will certainly impress the audience and help in the negotiations. Remember, in ancient China already powerful people would always wear gold or jade. No wonder that after real estate and cars, gold is extremely popular among wealthy Chinese.

竞争对手不是提供相似产品的品牌，而是提供相似好处的品牌。

我让你从这个令人震惊的消息中缓缓，让我们看看定位地图框架如何帮助识别所有这些意想不到的竞争对手并发现市场中的新机会。

图 7.4 遇见土豪（图画经由 DervalResearch 代戈提供，作者：Vlad Kolarov）
Figure 7.4 Meet the Tuhao 土豪 (Drawing courtesy of DervalResearch 代戈 , artist: Vlad Kolarov)

Competitors are not brands providing similar products but brands providing similar benefits.

I let you recover a bit from this shocking news and let's see how the positioning map framework can help identify all these unexpected competitors and detect new opportunities in the market.

7.3 定位：奈斯派索（Nespresso）案例 Positioning: The Nespresso case

定位能帮助我们明确自身产品在满足顾客需求方面不同于市场中其他产品的优势。让我们分析一下奈斯派索（Nespresso），看看它是如何处理这一难题的。

领先的企业总能游刃有余地运用这一方法，可惜留下的文字记录屈指可数。定位还能帮助我们发现新的机会。事实上也正是通过定位，我们才知道要开发哪类产品。这就是为什么奢侈品品牌在创建、革新或者扩张时尽快定位显得尤为重要的原因。

以奈斯派索（Nespresso）咖啡机为例，由于奈斯派索（Nespresso）的咖啡豆烘焙程度更深，品牌定位就是比星巴克（Starbucks） Via 速溶咖啡稍苦一点的口味。他们的竞争者可以是一杯好茶，或者你想保持清醒的话为什么不来一罐红牛呢。尽管就技术层面而言，奈斯派索（Nespresso）咖啡机和沁心浓（Senseo）或克里格（Keurig）之类的品牌类似，但是他们的优势和苦味程度不同，目标顾客群体也就不一样了。总之，即使公司提供相似的产品，如果针对的群体不同，它们也不会是各自的竞争对手。

Positioning will help us clarify the benefits of our offering compared to other products available in the market and regarding the customers' needs. Let's see with the Nespresso case how to approach that tricky subject.

This method is used by successful leading companies but was never documented so much. Positioning will also help us identify new opportunities. Actually, it is positioning that tells us what type of product to develop. It is why it is extremely important to plan this activity as soon as possible when creating, revamping, or expanding a luxury brand.

In the case of the Nespresso coffee machine, the positioning of a Nespresso cup is slightly more bitter than a Starbucks Via instant coffee cup because the coffee is more roasted. Competitors are also a good cup of tea, or why not a can of Redbull if you want to stay awake. Even if technically the Nespresso machine is similar to another single-serve device from brands like Senseo or Keurig, as it is not the same strength and bitterness, it doesn't target the same people. So, all the companies who are making similar products are not necessarily competitors because often they do not talk to the same type of people.

图 7.5 方方为 DervalResearch 代戈绘制的插画
Figure 7.5 Illustration by FangFang for DervalResearch 代戈

真正的竞争对手可能是生产其他产品的公司。

如果消费者购买奈斯派索（Nespresso）送礼的话，那么定位图就会变得完全不一样。

对于珠宝或手表等许多奢侈品配饰，以及智能手机或平板电脑等电子设备来说，很大一部分销量是作为礼品出售。

如果我们可以将实用的礼品与无用或装饰用的礼品区分开来，也可以将能彰显接受者身份的礼品与普通礼品区分开来，比如一部玫瑰金苹果手机（iPhone）。品牌想要成功定位的关键是明白购买产品或服务的角色是什么样子的，以及他们是出于何种目的。

我的建议是关注角色。

描述目标角色并了解他们的需求和动机以及购买产品时所处的境况将帮助我们确定真正的竞争对手。

例如，家里考虑购买时髦的奈斯派索（Nespresso）Citiz 咖啡机作为情人节礼物送给老婆，这样他们就能一起品尝美味的咖啡。可是转念一想，他又觉得一部玫瑰金苹果手机（iPhone）或一件精美的蒂芙尼（Tiffany）珠宝可能会更浪漫，更能讨老婆的欢心，说不定一高兴，老婆还会同意让他在家多玩玩 Xbox——平时家中的游戏机使用权可都掌握在女汉子老婆手里！

在另一种情况下，比如圣诞节礼物，辛迪就会考虑给男朋友买奈斯派索（Nespresso）的咖啡机，但又想到他可能会更喜欢飞利浦（Philips）的最新款剃须刀。说到飞利浦的男士剃须刀，礼品馈赠占了总销量的 40%，圣诞节或父亲节的销量最高。

So, the real competitors might be doing other types of products.

This is even more true if Nespresso is bought as a gift, the positioning map would then be completely different.

For many luxury accessories like jewelry or watches, as well as small consumer electronics, like smartphones or tablets, gifting represents a huge part of the sales.

We could for instance distinguish useful gifts from silly or decorative ones. Also, we can separate gifts that will provide the receiver with status, like a rose gold iPhone, from regular gifts. The key to successful positioning is to understand what persona will buy the product or service and for which purpose.

My advice is to focus on the persona.

Describing the target persona and understanding their needs and motivations as well as the context of the purchase will help us identify the real competitors.

Gary for instance is thinking of offering a trendy Nespresso Citiz to his wife for Valentine's Day, so they can enjoy a nice cup together. On the other hand, he thinks maybe a rose gold iPhone or a nice piece of Tiffany jewelry would be perceived as more romantic and give him more Xbox playing credits – as his "nuhanzi" wife, wearing the pants, is monitoring the access and usage of the gaming consoles in the house!

In another context, like Xmas' Day for instance, Cindy is wondering whether she should offer her boyfriend a Nespresso but also thinks he might enjoy the latest Philips Shaver. In the case of Philips shaving devices for men, even 40% of the sales are due to gifting with a peak at Christmas and on Father's Day.

要知道大部分购买者都是妻子和女朋友，吸引她们才是品牌提升销量的关键。于是，飞利浦在广告中甚至把剃须刀比喻成珠宝，宣称男性对他家剃须刀的钟爱就如同女性热爱钻石一样：广告语“就像你收到这个蓝色的小盒子一样”直白地暗示着著名的婚戒制造商蒂芙尼（Tiffany & Co.）。

请勿局限于同一产品类别或具有某些相同特征的产品。相反请关注产品会给消费者带来的好处。例如，奈斯派索（Nespresso）咖啡机除了能煮咖啡之外，还能因其炫酷的外形和潮流让消费者感到自己的地位。因此其他任何一款能彰显消费者地位的产品，如玫瑰金苹果手机（iPhone）都是可能的替代产品 (Derval, 2016)。

当分析这些方面的时候，定位图中每条坐标轴的标签就会自动出现。如果我们把奈斯派索（Nespresso）看作礼品，那么定位图描述的便是礼品类别，不是咖啡机市场。将礼物区分开来的两个可能维度是“功能”——礼物可能更有用或更无用——和“地位”，因为有些礼物会让你想在微信上分享照片，而另一些则不太想。

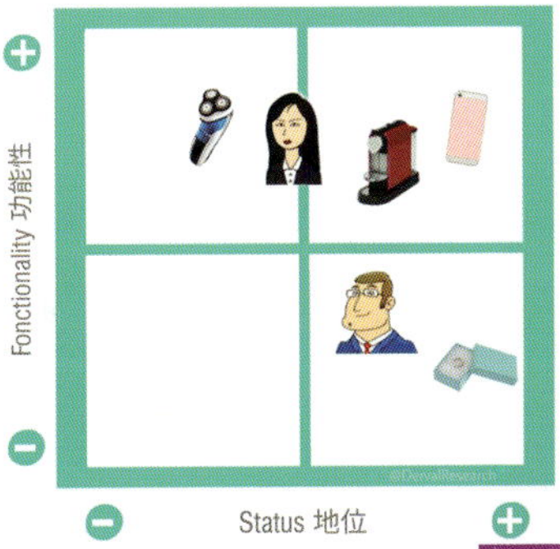

图 7.6 奈斯派索（Nespresso）的礼品定位图（经由 DervalResearch 代戈提供）

Figure 7.6 Nespresso positioning map on gifting (courtesy of DervalResearch 代戈)

Knowing that most gifters are indeed spouses and girlfriends, being appealing to women is key. And the brand even developed a campaign using an analogy with jewelry, explaining a Philips shaver is to the beloved one what a diamond is to a woman: “It's like you getting this little blue box”, the catch line is a direct reference to popular engagement ring maker Tiffany & Co.

Make sure not to limit yourself to the same product category or certain product features. Focus instead on the benefits the product will procure. For instance, a Nespresso machine can do coffee indeed but it will also provide the user with status as it is cool and trendy to own this machine. So, any other item that can provide status – like a rose gold iPhone – is a possible substitution product (Derval, 2016).

While analyzing all these aspects, the labels to use on each dimension of the positioning map will automatically emerge. If we consider Nespresso as a gift then the positioning map is describing the gifting category and not the coffee maker's market. Two possible dimensions that would set gifts apart are “functional” – as a gift might be more or less useful – and “status” as some gifts will make you want to share pictures on wechat and others less.

7.4 戴森的设计与创新
Design and Innovation with Dyson

价格永远不是问题，我重复，价格永远不是问题。

问题通常在于客户用所要求的预算换取所感知到的产品收益或增值。增加感知价值将使议题自动地转移到价格以外的其他领域，从而创造广阔的机遇（蓝海战略）。想想戴森以 10 倍的价格出售吹风机。

戴森使用无袋吸尘器，无螺旋桨风扇或镀金的吹风机，重塑了家用电器类别，不再让人梦寐以求。该公司由詹姆斯·戴森（James Dyson）在马姆斯伯里（Malmesbury）创建，现在也位于新加坡，营业额超过 40 亿欧元，其中四分之三的销售额来自亚洲。尽管戴森出售的产品价格更贵，但是戴森通过瞄准角色，展示其非常规设计产品的优势，坚持解决现实生活中的问题，还是神奇地打赢了这场巡回赛。

对于其所有类别，戴森提出以下三个问题：

a）我的产品能解决问题吗？

b）我的产品是否比替代产品更好地解决了问题？

c）对某些角色来说，该产品是否是必需品？

该品牌目前正在应对汽车市场，因为詹姆斯·戴森发现它效率低下，污染严重。

戴森打算用自己的汽车立即解决汽车的不良工程问题（如雨刷失效）和健康危害（柴油、安全）。

图 7.7 Aubery 为 DervalResearch 代戈绘制的插画
Figure 7.7 Illustration by Aubery for DervalResearch 代戈

Price is never the problem I repeat, price is never the problem.
The issue is often the benefits or added value of the product perceived by the client in exchange of the requested budget. Increasing the perceived value will automatically move the discussion to another field than price, creating a blue ocean of opportunities. Think of Dyson selling hair dryers for 10 times the price.
With bagless vacuum cleaners, propellerless fans, or gold-plated hair dryers, Dyson has reinvented the household appliances category, nobody was dreaming about anymore. The company created by James Dyson in Malmesbury, and now also based in Singapore, went past 4 billion euros turnover, with three quarters of the sales coming from Asia. Dyson managed this tour de force in spite of selling more expensive products, by targeting its personas, demonstrating the benefits of its unusually designed products, and sticking to solving real-life problems.
For all its categories, Dyson asks following three questions:
a) Does my product solve a problem?
b) Does my product solve the problem better than substitution products?
c) Is the product a must have for some persona?
The brand is currently tackling the automotive market, as James Dyson finds it ineffective and polluting.
Dyson intends to solve at once automotive bad engineering (like ineffective wipers) and health hazards (diesel, safety) with its own vehicle.

由于戴森是飞机迷，他承诺将在2021年发布这款电动汽车的彻底设计变革。目前还不清楚“戴森车”是在滚动还是在飞行——根据最新的专利申请，它至少应该具备一些越野能力。

但戴森决定重新把整个31亿美元的汽车项目的重点放在设计一个突破性的电池，而不是一个整车。该团队将发挥自己的优势，因为矛盾的是，自动驾驶汽车的主要问题是电池缺乏自主性。这都是因为电池讨厌温度的变化。这可能会导致众所周知的过热问题，但最近也发现了对寒冷副作用的厌恶，即电动汽车在 -7℃的温度下会失去一半的容量。这是戴森再次释放空气倍增技术的绝佳场所，因为电池可能是新的黑金。

品牌通过为目标角色设计必须拥有的产品而获益，这意味着价格不再是问题 (Derval, 2020)。

戴森（Dyson）总是着眼于同类产品不能让用户满意的地方，例如，戴森（Dyson）发现电子烘手机不太好用，会消耗大量能源，而且纸巾也不环保，这些都是问题。同样，真空吸尘器会因堵塞不能正常工作。

戴森（Dyson）生产了5216台样机后才找到上述三个问题的肯定答案，推出了第一代无尘袋真空吸尘器，可是分销商不明白这有何创新之处，又或许他们不想放弃有袋式真空吸尘器的利润。正是因为这个原因，无尘袋真空吸尘器在沉寂了14年后才得以问世：真是对品牌决心的一大考验。如果你有一个很棒的想法，即使有人嘲笑你，也不要轻易放弃！

As Dyson is a fan of planes, he promised a radical design change for this electrical vehicle to be released in 2021. It was not clear whether the “DysonMobile” would be rolling or flying – based on latest filled patents it should at least have had some off road capabilities.

But Dyson decided to refocus the whole $3.1 billion automotive program on designing just a groundbreaking battery instead of a whole vehicle. The team will play to its strength as paradoxically, the major issue with autonomous vehicles is the lack of autonomy of the battery. And this is all due to the fact that batteries hate variations in temperature. That can translate into well-known overheating issues but also recently discovered aversion to cold side-effects, where electric vehicles would lose half of their capacity at a temperature of -7°C. A perfect playground for Dyson to unleash once again their air multiplying technology in a very promising position as batteries might be the new black gold.

Brands gain at designing products that are must-haves for the target personas, meaning the benefits are such that the price is not a problem anymore (Derval, 2020).

Dyson always starts with users’ frustrations. For instance, Dyson saw that the electric hand dryer would not work very well, would consume a lot of energy, and the paper was not environmentally friendly, which is an issue. The same for the vacuum cleaner that was getting stuck and didn’t vacuum properly.

It took Dyson 5126 prototypes before being able to answer yes to these three questions and to launch the first version of his vacuum cleaner without the bag. Distributors didn’t get the innovation. Or maybe they didn’t want to give up the revenues generated by vacuum cleaners with bags. The product would take 14 years to hit the shelves for this reason: a big lesson of determination. If you have a great idea, never give up! Especially if people are mocking you.

7.5 新酒店设计 New Hotel Designs

年轻有才华的室内和景观设计师 Bonnie 和 TL 分享了他们关于如何在酒店这样复杂的空间中实现成功心流的方法。他们最近与 DervalResearch 代戈合作为领先和快速增长的酒店集团提供了 10 种独特的新酒店设计概念。在上海商学院、建桥学院、东华大学、四川大学和上海大学毕业生的宝贵投入下，他们在 6 周内、并在新冠病毒的第一波浪潮中完成了这项艰巨的任务！他们愿意分享关于如何创造室内和景观设计的 5 个步骤的宝贵想法：

第一步，了解需求。

“了解甲方需求，例如功能、设计要求，规划意图是这个过程棘手的部分，” Bonnie 透露，“我也了解到室内设计和景观设计之间存在巨大差异”。

Bonnie 解释了设计中对平衡的需求：“当我认识到一个人每天有超过 80% 的时间呆在建筑里时，我想的是如何去帮助他们更舒适地在空间中感受一切，并能解决一些现有的空间痛点。人们在建筑里居住、工作、学习、社交等，而室内设计师能做的就是让这些空间满足他们的功能需求，并且能够让他们在日复一日的生活中感受美好，而不是单调沉闷。”

第二步，找到一个主题。

Bonnie 的建议是理解客户给出的空间的功能、设计要求和规划意图，“想想空间的一些形容词，比如怎样的居住空间、怎样的办公空间。这样更容易让我得到想要的理想方案”。

Bonnie and TL, young talented interior and landscape designers, share their approach on how to achieve a successful flow in a complex space like a hotel. Their recent collaboration with DervalResearch 代戈 led to the creation and rendering of 10 unique new hotel design concepts for a leading and fast-growing hospitality group. They completed this herculean task – with the precious input of Shanghai Business School, Jian Qiao, Donghua, Sichuan and Shanghai University graduates – within 6 weeks and in the midst of the coronavirus first wave! They accepted to share their nuggets of wisdom on how to create interior and landscape designs in 5 steps:

Step 1. Understanding the needs.

“Understanding the needs of Party A, such as functions, design requirements, plan intentions is the tricky part of the process” confides TL “and I also learned that there is a huge difference between interior and landscape design.”

Bonnie explains the need for balance in the design: “When I realized that a person spends more than 80% of the time in a building, every day, I thought about how to solve some of the existing pain points and help them feel more comfortable in that space. People live, work, study, socialize in the building, and what interior designers can do is to make these spaces meet their functional needs, make them feel good in their daily lives, but without making the spaces dull.”

Step 2. Finding a theme.

Bonnie’s tip to understand the functions, design requirements, and program intentions of the space given by the client is to “think of some adjectives for the space, such as what kind of living space or what kind of office space it is. This makes it easier for me to get the ideal plan I want”.

“就像这次在戴安娜教授的指导下我们做的十套酒店方案一样，我们首先定义一个空间的主题，再用几个形容词给这个空间一个初步的概念，以及关联的品牌代码”。

第三步，呈现概念。

根据概念方案（甲方修改意见），提出进一步的方案，汇报（从平面转向立体空间，建模，渲染，利用工具 SKETCHUP、ENSCAPE、LUMION 等）：

- 在设计材料、设计方案初步敲定好后用 AUTOCAD 画出平面与立面图。
- 画完的平面图导入 SKETCHUP 建模软件进行建模、贴材质。
- 完成模型后用 SKETCHUP 的 VRAY 渲染插件进行空间的渲染，或者将模型导入 LUMION 渲染器进行渲染，完成效果图。

第四步，让设计焕然一新。

深化设计与一些细节处理，如配件，有时还有人（或机器人），这将有助于让设计变得生动起来，正如你在特色酒店设计中看到的那样。

第五步，完成漂亮的设计。

TL 强调了设计中美感的重要性：“意识到基本上每个人全天时间都花在工作、学习这种高压力状态下时，我们如何去缓解这种压力呢？我相信都会有对‘美’的向往。景观是人所向往的自然，景观是人类的栖居地，景观是人造的工艺品，景观是需要科学分析方能被理解的物质系统，景观是有待解决的问题，景观是可以带来财富的资源，景观是反映社会伦理、道德和价值观念的意识形态，景观是历史，景观是美。”

“Just like the ten hotel plans we made with DervalResearch 代戈, we first defined a space theme, and then used a few adjectives to give the space a preliminary concept, and associated brand codes”.

Step 3. Rendering the concept.

According to the conceptual plan and revised opinions, further plans are proposed and rendered using:

- AUTOCAD to draw the plane and elevation after preliminary finalization of the design materials and design plan.
- SKETCHUP modeling software for modeling and attaching materials to the floor plan.
- SKETCHUP's VRAY rendering plug-in and LUMION to finalize the rendering.

Step 4. Bringing the design to life.

Deepening the design with some detail processing like accessories and sometimes people (or robots) will help bring it to life, as you can see it in the featured hotel designs.

Step 5. Finalizing the beautiful design.

TL highlights the importance of beauty in the design: “When you realize that basically everyone spends the entire day working or studying under great pressure, the question is how can we relieve this pressure? I believe there will always be a yearning for ‘beauty’. The landscape is the nature that people yearn for. The landscape is the habitat of human beings. The landscape is a man-made handicraft. The landscape is a material system that needs scientific analysis to be understood. The landscape is a problem to be solved. The landscape is a resource that can bring wealth. Landscape is an ideology that reflects social ethics, morals and values. Landscape is history and landscape is beauty.”

图 7.8 GAMING 电竞房 (经 DervalResearch 代戈许可印刷)
Figure 7.8 GAMING esports pad (printed with DervalResearch 代戈 permission)

图 7.9 GAMING 篮球场 (经 DervalResearch 代戈许可印刷)
Figure 7.9 GAMING basketball court (printed with DervalResearch 代戈 permission)

GAMING concept: 余 卧 龙 （ Jordan ）, 上海建桥学院 , jewelry design; 蒲玮（Vivian）, 上海建桥学院 , business; 刘晨（Mark), 上海建桥学院 , luxury; 陈昊（Ryan）, 上海建桥学院 , luxury; 谭亮 (TL) , 上海商学院 , landscape architecture
Design and rendering: 谭亮 (TL) , 上海商学院 , landscape architecture

图 7.10 LODGE 小屋 (经 DervalResearch 代戈许可印刷)
Figure 7.10 LODGE cabin room (printed with DervalResearch 代戈 permission)

图 7.11 LODGE 大空间水疗房 (经 DervalResearch 代戈许可印刷)
Figure 7.11 LODGE infinity spa (printed with DervalResearch 代戈 permission)

LODGE concept: 边瑾瑜 (Anita), 四川农业大学； 余鹰鸽 (Rima), 四川农业大学 , 市场营销系； 秦思伟 (Bonnie), 上海商学院 , landscape architecture
Design and rendering: 秦思伟 (Bonnie), 上海商学院 , landscape

7

7.6 Gucci 如何赢得中国消费者的心 How Gucci won China over

在这场争夺中国消费者的巡回赛中，古驰 (Gucci) 神奇地将路易威登（LV）与爱马仕（Hermès）击败，在中国消费者心中（和皮带上）赢了一场刻骨铭心的胜仗。古驰，在首席执行官马可・比扎里（Marco Bizzarri）领导的开云品牌营业额翻了一番，并产生了数百万次在线搜索。然而古驰 (Gucci) 是如何做到如此成功的？其他针对中国市场的品牌可以从中学到什么？

古驰（Gucci）通过款式、剪裁，以及跨性别的方法，并用“中国优先”这一高招儿，搞定了中国消费者。 如今，该品牌借助沉香和黄金，正忙着将其成功的秘诀出口到其他地区，例如中东和非洲。

马可・比扎里（Marco Bizzarri）在从斯特拉・麦卡特尼（Stella McCartney）品牌卸任继而入驻古驰 (Gucci) 后，决定用动物图案替代毛皮。古驰（Gucci）使用的动物和花卉图案对中国消费者非常有吸引力。 古驰（Gucci）新任设计师 Michele Alessandro 通过生动的色彩和花卉图案，成功地抓住了中国人的审美价值。 问题是中国市场现在正在推动全球奢侈品消费，因此无论中国消费者喜欢什么，什么便将成为潮流。与此同时，千禧一代渴求真实性，古驰（Gucci）带来的奢华设计，他们很容易认同，进而使得他们与众不同。

图 7.12 方方为 DervalResearch 代戈绘制的插画
Figure 7.12 Illustration by FangFang for DervalResearch 代戈

Gucci managed the tour de force to dethrone LV and Hermès in the heart and on the belts of Chinese consumers. The Kering brand led by CEO Marco Bizzarri, doubled its turnover and generates millions of online searches. But how did Gucci make it? What can other brands targeting the Chinese market learn from it?

Gucci nailed it with Chinese shoppers from the patterns and fit to the cross-gender approach, and the brilliant idea to put China first. The brand is now busy exporting their successful recipe to other regions like the Middle East and Africa, with the help of oud and gold.

Marco Bizzarri, when shifting from Stella McCartney to Gucci, decided to replace fur with animal prints. The animal and floral prints used by Gucci are very appealing to Chinese customers. New Gucci designer Michele Alessandro managed to capture the aesthetic values of the Chinese, through dynamic colors and floral prints. Thing is that the Chinese market is driving the global luxury consumption now, so whatever the Chinese like, becomes the trend. Also, millennials are looking for authenticity and Gucci comes with very extravagant designs they can easily identify with, and that make them stand out.

古驰（Gucci）深知中国市场蕴含的力量既可以迎合潮流，又可以拒绝潮流。

香奈儿(Chanel)和爱马仕(Hermès)似乎在国际范围内提供的货物品种相同，但是仔细观察后不难发现，上海的香奈儿（Chanel）店比巴黎旗舰店参考并展示了更多闪亮的珠宝 (Derval, 2020)。

古驰（Gucci）更进一步，不但在本土市场提供迎合中国购物者的货物品种，而且在全球市场范围内迎合中国购物者。此举颇有成效，古驰（Gucci）的销售额猛增，占开云集团（Kering Group）收入的三分之二。有些产品显然基于中国文化和品味，随后在全球范围内推广。

图 7.13 方方为 DervalResearch 代戈绘制的插画
Figure 7.13 Illustration by FangFang for DervalResearch 代戈

Gucci understood indeed that China had the power to adopt or reject trends.
Chanel and Hermès seem to propose the same assortment worldwide but when you have a closer look you can notice that for instance the Chanel stores in Shanghai will reference and display more shiny jewelry than the flagship store in Paris (Derval, 2020).
Gucci even went a step further by globally adapting the products for Chinese shoppers instead of just adapting their local assortment. And it paid off as Gucci sales have skyrocketed to represent two thirds of Kering Group revenues. Some products were clearly rooted in Chinese culture and taste and rolled-out at a global level.

合身度对品牌来说可谓是成也萧何败也萧何。例如，杰尼亚（Ermenegildo Zegna）利用生理事实在中国建立了一个量身定制西装的帝国：西方男人的手臂通常比东方男人长，因此意大利西装对中国男人来说普遍袖子太长。效仿这一观测，普拉达（Prada）当然还有古驰（Gucci）开始提出在产品中使用 ¾ 袖。合身度的重要性不仅体现在服装，对配饰也同样至关重要，诸如古驰（Gucci）之类的品牌专门针对亚洲市场研发的墨镜，会有特殊的鼻梁款式和墨镜前面的特殊设计。

尽管你在古驰（Gucci）网站上仍然可以找得到“男性”和“女性”条目，但该品牌性别之间的界限是模糊的。

尤其在上海这样既有女汉子(强势,豪放又有女人味的女人),又有小男人(字面上的“小丈夫”,指时尚,花很多时间打扮的男人)的地区。如果衣服太过夸张,时尚达人可以配一瓶例如古驰罪爱沉香（Gucci Guilty Oud）这样的中性香水,或是有兽性的配饰譬如 Gucci 带有蜜蜂、蛇或老虎刺绣的手表,这些都是大胆的中国购物者的不二之选,鉴于他们已经敢于购买顶级的比特犬 T 恤衫和手枪包。

古驰（Gucci）接下来要做什么？古驰罪爱沉香（Gucci Guilty Oud）香水的推出被视为该品牌向中东以及非洲市场方向发展的典型举措。沉香木，又名“众神之木”，价格高昂堪比黄金。通过此举试图改变中国消费者将会是一个蛮有趣的论点，尽管中国消费者更倾向于淡雅委婉的香氛，例如高田贤三的一枝花香水（Flower by Kenzo）。

Fit can make or break a brand. Ermenegildo Zegna for instance took advantage of a physiological fact to build a tailor-made suit empire in China: Western men usually have longer arms than Eastern men and as a result Italian suits would have too long sleeves for Chinese men. Learning from this observation, Prada and of course Gucci started proposing collections with ¾ sleeves. Fit is critical not only for clothes but also for accessories, and brands like Gucci develop products specifically for the Asian market, with a special nose bridge style, and specific designs for the front of the sunglasses.

Even though you still find a “Men” and a “Women” entry on the Gucci website, the borders between genders are blurred. This is especially true in areas like Shanghai with local nǚhànzi (powerful women who are fierce and feminine) and xiǎo nánrén (literally “little husband”, and stylish men who spend more time grooming). And if clothes are too bold, fashionistas can grab a unisex fragrance like Gucci Guilty Oud or beasty accessories like Gucci watches with an embroidered bee, snake, or tiger, spot on for our fierce Chinese shoppers, who already dare over the top pitbull t-shirts or gun holder clutches.

What is next for Gucci? The launch of Gucci Guilty Oud was seen as a typical move in direction to the Middle Eastern market but also Africa. Oud, known as agarwood or “the wood of gods”, can be as expensive as gold. That could be an interesting argument to try to convert Chinese too, even though they tend to prefer softer scents like Flower by Kenzo.

7.7 练习 Exercises

7A 月球酒店

为月球上的新酒店设计一个卧室和一个共享房间，要考虑到当地的条件。解释一下这个心流将如何完美地适合你的目标角色。

7B 中东的保时捷案例

品牌经理在打造或进行品牌扩张时倾向于马上分析竞争对手，其实这往往是一种误导。因为首先竞争对手的做法不一定准确，其次，我们可能没有看到真正的竞争对手，而这点尤其重要。

以保时捷（Porsche）卡宴为例，4x4 车型在迪拜非常流行，那它的竞争对手是谁呢?

您可以在 DervalResearch 代戈微信上获得定位图框架 (图 7.15): 扫描二维码并键入关键词 **设计 7**。

图 7.14 方方为 DervalResearch 代戈绘制的插画
Figure 7.14 Illustration by FangFang for DervalResearch 代戈

7A A Hotel on the Moon

Design a bedroom and a shared space for a new Hotel on the Moon, taking into account local conditions. Explain how the flow will be perfect for your target persona.

7B The Porsche case in the Middle East

When creating or expanding a brand, managers tend to immediately look at competitors. It is often misleading because a) competitors might be doing silly things and most importantly b) we might not be looking at the real competitors.

Let's take the example of the Porsche Cayenne, this 4x4 model is very popular in Dubai. But who are the competitors?

You can get the positioning map framework (Figure 7.15) on DervalResearch 代戈 wechat: scan the QRcode and type the keyword design7.

7B 答案

法拉利（Ferrari）？路虎（Land Rover）？还是宝马（BMW）？你离答案还远着呢。

保时捷卡宴经常作为第二辆车来购买。它不是用来从 A 到 B，而是用来和朋友一起享受闲暇时光或炫耀一下。因此，根据用途，游艇、海滩别墅或一件漂亮的珠宝就可以了。

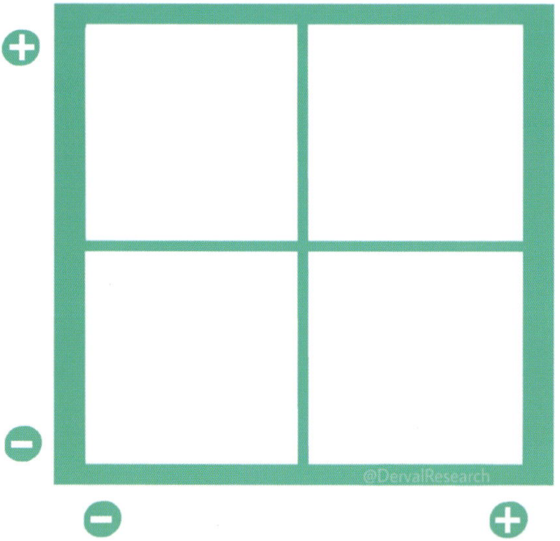

图 7.15 定位图框架（经由 DervalResearch 代戈提供）
Figure 7.15 Positioning Map framework (courtesy of DervalResearch 代戈)

Solution to 7B

Ferrari? Land Rover? BMW? You are nowhere near the answer.

The Porsche Cayenne is often bought as a second car. It is not used to go from A to B, but to enjoy leisure time with friends or show off a little bit. So, depending on the purpose, a yacht, a villa on the beach, or a nice piece of jewelry will do.

一只熊，一只猴子和一个香蕉结果

a）熊和猴子 | 香蕉

毫不犹豫地把熊和猴子分成一组的人，因为这两个都是动物，而另一边是水果，往往具有更高密度的多巴胺 D2 受体。他们将信息存储在类别或框中，并以顺序或线性方式思考。新颖性是具有破坏性的，因为它有时需要创建新的类别。另一方面，分析思维是快速和有效的，因为所有需要的数据都已经分组了。这个心流应该有层次结构，并有明确的类别——他们倾向于在鞋店买鞋子。

b）猴子和熊 | 香蕉，或 猴子和香蕉 | 熊

在将猴子与熊或香蕉分组之间犹豫不决的人往往具有平均密度的多巴胺 D2 受体。他们将信息存储到类别或框中，但同时在其他关联之间创建索引。因此，他们既可以顺序思考又可以类比思考。他们是跨职能团队中的关键，以便在通常不理解彼此的非常顺序和非常类比的人员之间翻译沟通。只要有人与人之间的互动空间，这种流动可以用多种方式构建——他们倾向于买推荐的鞋子。

c）猴子和香蕉 | 熊

毫不犹豫地把猴子和香蕉分成一组的人，因为一个吃了另一个，往往具有较低密度的多巴胺 D2 受体。他们将信息与其上下文和属性存储在一起。例如，动物是与颜色或饮食有相同水平的属性。每个属性都是一个索引，并与其他属性相链接，这有助于从多个角度以及与上下文一起检索数据。人们可以很容易地通过类比来比较情况并得出新的想法。这个心流应该像一个思维导图一样，以一种视觉化的方式包含许多可能的条目——他们不介意在同一个地方买鞋和巧克力。

The Bear, the Monkey, and the Banana Results

a) Bear and Monkey | Banana

People grouping without any hesitation the bear with the monkey, because these two are animals, and fruits on the other side tend to have a higher density of dopamine D2 receptors. They store information into categories or boxes, and think in a sequential or linear way. Novelty is disruptive as sometimes it requires the creation of new categories. On the other hand, analytical thinking is fast and effective as all the needed data is already grouped. The flow should be structured hierarchically and with clear categories - they tend to buy shoes in a shoe store.

b) Monkey and Bear | Banana, or Monkey and Banana | Bear

People hesitating between grouping the monkey with the bear or with the banana tend to have an average density of dopamine D2 receptors. They store information into categories or boxes, but at the same time create an index between other associations. They can therefore think both in a sequential and in an analogical way. They are key in cross-functional teams in order to translate communications between very sequential and very analogical people who usually do not understand each other. The flow can be structured in multiple ways as long as there is room for human interaction – they tend to buy shoes that are recommended.

c) Monkey and Banana | Bear

People grouping without any hesitation the monkey with the banana, because one eats the other, tend to have a lower density of dopamine D2 receptors. They store information together with their context and attributes. For instance, animal is an attribute at the same level as the color or the diet. Each attribute is an index and is linked to the others, which helps retrieve the data from many angles and together with the context. People can easily compare situations and come to new ideas by analogy. The flow should be structured like a mindmap, in a visual way with many possible entries – they do not mind buying shoes and chocolate at the same place.

7.8 要点 Take-Aways

- 要考虑中国消费者的感官偏好（颜色，形状，图案，跨性别）以及体态特征（合适的眼镜，臂长，身材，肤色）
- 调整分类，以便当地商店可以提供合适的商品
- 接受深度调整品牌代码，以适应当地口味，并在需要时借助品牌合作
- 通过为目标角色设计必要的产品和体验来改变市场游戏规则
- 确定真正的竞争对手，他们通常是提出截然不同的产品和服务，但却着眼于同一个角色的公司

- Take into account shoppers' sensory preferences (colors, shapes, patterns, cross-gender) and morphology (fitted glasses, arm length, body shape, skin tones)
- Adapt the assortment, so that local stores can propose suitable items
- Accept to deeply adapt the brand codes to fit local taste, with the help of brand collaborations when needed
- Change the game in the market by designing products and experiences that are must-haves for the target personas
- Identify the real competitors, who are often companies proposing very different products and services but having their eyes on the same persona

7.9 术语表 Glossary

品牌合作

“品牌合作是一种超越产品的战略，可以赋予员工、消费者和公民意义，既包括与零售商、非政府组织、地区的合作，也包括与价值观甚或社会事业相关的名人的合作。” (Michel & Willing (2020)

等待营销

“等待营销是指在合适的时间、在正确的地点、当人们接受时进行沟通。” (Derval, 2018)

心流

“在当下着手的某件事情或某个目标中，全神贯注、全情投入并享受其中而体验到的一种精神状态。” (Nakamura& Csikszentmihalyi, 2009)

Brand Collaboration

“Brand collaboration is a strategy that goes beyond the product and can give meaning to employees, consumers and citizens, involving both collaborations with retailers, NGOs, territories and also collaborations with celebrities that are associated with values or even social causes.” (Michel & Willing, 2020)

Wait Marketing

“Wait Marketing is about communicating at the right moment at the right place, when people are receptive.” (Derval, 2018)

Flow

“Flow refers to a state of mind that people experience when they are immersed in something or a goal that they are working on at the moment, fully absorbed, fully engaged, and enjoying it.” (Nakamura& Csikszentmihalyi, 2009)

7.10 参考文献 References

Derval, D. (2022). The Right Sensory Mix: Decoding Customers' Behavior and Preferences. Springer Nature, ISBN: 978-3-662-63794-4.

Derval, D. (2020, December 15). سر انجذاب الصينيين إلى ماركة غوتشي. Harvard Business Review Arabia. Retrieved from https://hbrarabic.com

Derval, D. (2019) The Dyson Multiplying Effect. NUST Business Review, Vol 1, Issue 1, DOI: https://doi.org/10.37435/NBR-19-0108

Derval, D. (2018). Designing Luxury Brands: The Science of Pleasing Customers' Senses. Springer Business & Management, ISBN: 978-3-319-71555-1.

戴安娜 · 代尔瓦勒 (Diana Derval). 奢侈品品牌营销：创建 · 实施 · 案例 (Luxury Brand Marketing). 东华大学出版社 (2016), ISBN: 9787566910424.

Derval, D., Bremer, J. (2012). Hormones, Talents, and Career: Unlock your Hormonal Quotient. SpringerBriefs, ISBN: 978-3-642-25712-4.

Derval, D. (2011). Réussir son étude de marché en 5 jours. Eyrolles, ISBN: 978-2-212-54958-4.

DervalResearch (2020). The science of colors: From product to experience design. Iversity, Springer Nature Group. https://iversity.org/en/courses/the-science-of-colors

Michel, G., & Willing, R. (2020). The Art of Successful Brand Collaborations: Partnerships with Artists, Designers, Museums, Territories, Sports, Celebrities, Science, Good Cause… and More. Routledge.

Nakamura, J., & Csikszentmihalyi, M. (2009). Flow theory and research. Handbook of positive psychology, 195-206.

总结 Conclusion

感谢您阅读这本关于设计科学的书。我希望它能为您提供有用的工具和知识，以便：

- 了解人们的行为和感知
- 预测设计偏好
- 设计目标客户体验
- 定义吸引人的品牌代码
- 设计赏心悦目的产品
- 将产品设计和体验本地化

通过本书中分享的各种商业案例、访谈和见解，您可能已经感觉到设计实际上非常容易理解。现在，您将像 DervalResearch 代戈团队和我们的客户一样，具有独到的洞察力。

我们都是消费者，但首先是人，甚至在此之前是动物。我们的生理构成将决定我们在生活、行为和偏好中的动机。这就是为什么人们如此独特，同时又如此可预测的原因。了解行为和偏好背后的机制是设计的关键组成部分。当然，诸如品牌代码、感觉剖面、角色、动机、创新路线图、影响者地图和定位地图等工具将帮助您实施成功的设计策略，以创建、改造和发展成功的可持续品牌。

请确保与我联系，并带来你的设计和令人兴奋的项目的最新消息！

Thank you for reading this book on the science of Design! I hope it provided you with useful tools and knowledge to:

- Understand people's behavior and perception
- Predict design preferences
- Design targeted customer experiences
- Define appealing brand codes
- Design eye-pleasing products
- Adapt product designs and experiences locally

Through the various business cases, interviews, and insights shared in this book, you might have sensed that design is actually very simple to understand. Now. like DervalResearch 代戈 teams and our clients you will be able to see what others fail to see.

We are all consumers, but first people, and even before that, animals. Our physiological makeup will condition our motivation in life, behavior and preferences. This is the reason why people are so unique and at the same time so predictable. Understanding the mechanisms underlying behavior and preferences is the key component of design. Then, of course tools like brand codes, sensory profile, persona, motivations, innovation roadmap, influencers' map, and positioning map, will help you implement winning design strategies to create, revamp, and develop successful and sustainable brands.

Make sure to connect and to keep me posted on your designs and exciting projects!

戴安娜 Diana Derval

关于戴安娜·代尔瓦勒教授 About Prof. Diana Derval

戴安娜·代尔瓦勒教授，博士，EMBA，是 DervalResearch 代戈的主席兼研究总监，这家全球研究公司专门研究人类行为和偏好。哈佛商业评论撰稿人，CES 亚洲创新奖评审员，获得爱迪生奖提名的荷尔蒙商®（HQ）专利发明者，以及 1000 万人参加的代尔瓦勒色彩测试®专利发明者。戴安娜是行为神经内分泌学学会成员，著有《设计奢侈品品牌》、《奢侈品品牌营销》和《正确的感官组合》等书，是 Berry AMA 最佳营销书籍奖的入围者，由 Philip Kotler 推荐。戴安娜将引人入胜的神经科学突破转化为强大的商业框架识别、理解和预测人类的特征、动机和行为。戴安娜加速了财富 500 强企业的发展，包括丝芙兰、米其林、索菲特、飞利浦和欧莱雅。戴安娜·代尔瓦勒在索邦商学院、复旦 / 麻省理工学院和东华大学教授创新、奢侈品、设计思维和神经科学。从巴黎到上海，25000 多名专业人士都喜欢戴安娜的励志讲座、TEDx 讲座和研讨会。

欢迎您通过 linkedin 联系，发送电子邮件至 diana@dervalresearch.com 或微信。

Diana Derval, PhD, EMBA, is Chair and Research Director of DervalResearch, global research firm specializing in human behavior and preferences. Harvard Business Review contributor, Jury for CES Asia Innovation Awards, patented inventor of the Hormonal Quotient® (HQ) nominated for the Edison Awards and of the Derval Color Test® taken by 10 million people. Member of the Society for Behavioral Neuroendocrinology, and author of the books *Designing Luxury Brands*, *Luxury Brand Management* (《奢侈品品牌营销》), and *The Right Sensory Mix* – finalist of the Berry-AMA Award for best marketing book and recommended by Philip Kotler. Diana turns fascinating neuroscientific breakthroughs into powerful business frameworks to identify, understand, and predict human traits, motivations, and behavior. She has accelerated the development of Fortune 500 firms including Sephora, Michelin, Sofitel, Philips, and L'Oréal. Diana Derval teaches Innovation, Luxury, Design Thinking and Neurosciences at Sorbonne Business School, Fudan/MIT and Donghua University. Over 25,000 professionals enjoyed her inspirational lectures, TEDx talks, and workshops from Paris to Shanghai.
You are welcome to connect via linkedin, email to diana@dervalresearch.com or wechat.

同一个作者写的书 Books by the same author

《正确的感官组合》
Derval, D. (2022). The Right Sensory Mix: Decoding Customers' Behavior and Preferences. Springer Nature. 978-3-662-63794-4.
《色彩科学》
DervalResearch (2020). The Science of Colors: From product to experience design. Iversity, Springer Nature Group. https://iversity.org/en/courses/the-science-of-colors
《设计奢侈品品牌》
Derval, D. (2018). Designing Luxury Brands. Heidelberg: Springer Nature.
《奢侈品品牌营销》
Derval, D. (2016). Luxury Brand Management. Shanghai: Donghua University Press.
《荷尔蒙、天赋和职业》
Derval, D., Bremer, J. (2012). Hormones Talent and Career. Heidelberg: Springer.
《五天内成功完成你的市场调研》
Derval, D. (2011). Réussir son étude de marché en 5 jours. Paris: Eyrolles.
《等待 · 营销》
Derval, D. (2006). Wait Marketing: Communiquer au bon moment, au bon endroit. Paris: Eyrolles.

戴安娜 · 代尔瓦勒教授（图画经由 DervalResearch 代戈提供，作者：Vlad Kolarov）
Prof. Diana Derval (Drawing courtesy of DervalResearch 代戈 , artist: Vlad Kolarov)

关于 DervalResearch 代戈生物技术 About DervalResearch 代戈

DervalResearch 代戈是一家利用生物科学来解码人类行为和偏好的先锋研究公司。人们是理性的，他们只是有非常不同的感知。角色的决定似乎只是取决于文化和情感，但实际上是由深深植根于他们自己的生理、微生物群、感官感知和激素构成，因此可以预期和更好地针对。我们的科学家和设计师团队借助荷尔蒙商®(HQ)、感官地图®或代尔瓦勒色彩测试®等非侵入性但强大的预测工具，帮助组织抓住商业机会，提高创新命中率，开发从巴黎到上海的地球和人类友好的体验。

更多信息请访问 www.dervalresearch.com 或者我们的企业微信。

DervalResearch 代戈 is a pioneer research firm using biosciences to decode human behavior and preferences. People are rational, they just have a very different perception. Personas' decisions seem to just come down to culture and emotions but are in fact deeply rooted in their very own physiology, microbiome, sensory perception, and hormonal makeup, and can therefore be anticipated and better targeted. With non-invasive yet powerful predictive tools like the Hormonal Quotient® (HQ), Sensory GeoMaps®, or the Derval Color Test®, our neurodiverse team of scientists and designers help organizations seize business opportunities, increase their innovation hit rate, and develop planet and people-friendly experiences from Paris to Shanghai.
You will find more information at www.dervalresearch.com or on our corporate wechat.